NEW LINES

New Lines

CRITICAL GIS AND THE TROUBLE OF THE MAP

Matthew W. Wilson

University of Minnesota Press
Minneapolis
London

An earlier version of the Introduction was published as "New Lines? Enacting a Social History of GIS," *Canadian Geographer* 59, no. 1 (2015): 29–34; copyright 2014 by Canadian Association of Geographers / L'Association canadienne des géographes; reprinted courtesy of John Wiley and Sons, Inc. An earlier version of chapter 1 was published as "OXAV," in Ali Fard and Taraneh Meshkani, eds., *New Geographies 07* (Cambridge, Mass.: Harvard University Press, 2015), 174–77. An earlier version of chapter 4 was published as "Paying Attention, Digital Media, and Community-Based Critical GIS," *Cultural Geographies* 22, no. 1 (2015): 177–91; copyright 2015 by SAGE Publications. doi:10.1177/1474474014539249

Copyright 2017 by the Regents of the University of Minnesota

All rights reserved. No part of this publication may be reproduced, stored in a retrieval system, or transmitted, in any form or by any means, electronic, mechanical, photocopying, recording, or otherwise, without the prior written permission of the publisher.

Published by the University of Minnesota Press
111 Third Avenue South, Suite 290
Minneapolis, MN 55401-2520
http://www.upress.umn.edu

The University of Minnesota is an equal-opportunity educator and employer.

Library of Congress Cataloging-in-Publication Data
Names: Wilson, Matthew W., 1981– author.
Title: New lines : critical GIS and the trouble of the map / Matthew W. Wilson.
Description: Minneapolis : University of Minnesota Press, 2017. | Includes bibliographical references and index. |
Identifiers: LCCN 2017001744 (print) | ISBN 978-0-8166-9852-3 (hc) | ISBN 978-0-8166-9853-0 (pb)
Subjects: LCSH: Geographic information systems—Social aspects. | Digital mapping. | Digital maps. | BISAC: SOCIAL SCIENCE / Human Geography. | TECHNOLOGY & ENGINEERING / Cartography. | TECHNOLOGY & ENGINEERING / Social Aspects.
Classification: LCC G70.212 .W55 2017 (print) | DDC 910.285—dc23
LC record available at https://lccn.loc.gov/2017001744

CONTENTS

	Preface	vii
	Introduction: But Do You Actually *Do* GIS?	1
1	Criticality: The Urgency of Drawing and Tracing	25
2	Digitality: Origins, or the Stories We Tell Ourselves	47
3	Movement: Strange Concepts and the Essentially Subjective	69
4	Attention: Memory Support and the Care of Community	95
5	Quantification: Counting on Location-Aware Futures	115
6	A Single Point Does Not Form a Line	135
	Acknowledgments	143
	Notes	145
	Index	179

PREFACE

> Cartography is perhaps the chief tool-metaphor of technoscience.
> —DONNA HARAWAY, *MODEST_WITNESS@SECOND_MILLENNIUM*

> Wars. So many wars. Wars outside and inside.
> —BRUNO LATOUR, "WHY HAS CRITIQUE RUN OUT OF STEAM?"

Geographic information systems are more than what GIS users and developers tell us they are. This advice extends a position held by Brian Harley in 1989 regarding the field of cartography[1]—and this is precisely the kind of suspicious attention that propels much so-called critical thought toward technology today. *Do not trust that technicians might actually understand their social situation and conditions. There is more to their stories.* Is this not always the case? Harleian suspicion belies a dangerous implication—that the use of GIS runs counter to reasonable, situated, and radical knowledges. To call into question those stories told by users and developers of GIS effectively establishes an inside and an outside of critique, a war of either/or. But it is not productive to insist that you are either with us or against us.[2] Other paths are and have been possible. The purpose of what follows is to walk the path between practice and theory. The point is to trouble all too easy distinctions, not to resolve them, but to live them, to stay with the trouble, as Donna Haraway would insist:[3] cartography and GIS are made, not made up. Or to invoke Deleuze and Guattari, we must trust the conjunctive force of the rhizome: "and . . . and . . . and . . .": always addition, never subtraction.[4]

Mapping technologies occupy a curious status in society. Profoundly, they are *objects* of publicity, to translate the real world and our being within it—from the earliest TO maps and mappa mundi to the release of Google Earth in the web browser. At the same time, they are *objective*, the products of technicians, expertise, science. I prepared my thoughts for this preface while walking the labyrinth of the Gothic Quarter in Barcelona, Spain. Here, wanderers are reminded of the power of a most basic technocultural phenomenon: the reading of a map.[5] Captured within its medieval streets, visitors unfold paper maps in absence of their digital counterparts. GPS is highly undependable here, with narrow, winding paths and several-story stone buildings closing in what intimate public space exists. Visitors contort the map and their bodies to orient to this place. They trace with their fingers the path traveled. They gaze upward, past the gargoyles and bell towers, attempting to sort north from south, east from west, mountain from sea, largely in vain. In this moment, one can feel the pervasiveness of the location-aware society. The absence of digital devices can leave us feeling peculiarly vulnerable. For some, being lost has become a more acute sensation. How did we arrive here, now? What are the implications for these devices and practices that swaddle society in location-rich media, for these new lines that draw us in?

There are many entry points into this conversation: the mid-twentieth-century history of computing, the technological and financial speculation of the early 2000s, the devolvement of the state under neoliberal governance, and the rise of technolibertarians and post-9-11 security assemblages. However, the story I intend to weave encircles an ongoing discussion around digital mapping and GIS—the definitions of which have been sources of great indigestion. In 1999 Nick Chrisman sought to settle a definition of general acceptance for GIS. Synthesizing previous attempts, he arrived with concision at the following definition:

> Geographic Information System (GIS)—Organized activity by which people measure and represent geographic phenomena then transform these representations into other forms while interacting with social structures.[6]

Largely similar to earlier definitions (indeed, he was being synthetic!), Chrisman introduces a key departure—placing GIS in relation to social structures.[7] His new definition is particularly motivated by a commonly held earlier alternative from 1989:

> A system of hardware, software, data, people, organizations and institutional arrangements for collecting, storing, analyzing and disseminating information about areas of the earth.[8]

This earlier definition served to constitute the technology in atomistic ways. People and institutions were users and developers of this technology—part of the overall system, with discrete inputs, outputs, and relations. And while GIScientists no doubt believed that the technology was changing society, understanding that change fell outside the remit of their field, into the void of "cultural" inquiry. These changes could not possibly impact tool development and use.

Despite these beliefs, society changed. New technocultural relationships were forged, and the field of GIScience scrambled to figure out what was happening. Journal publications and conference sessions on "web GIS" were abundant just as GIScientists stood in line at Apple stores to get the first iPhone and to experience using digital maps *in situ*. More than a disconnect, this seemed to be a chasm. We were ill equipped. I was in the middle of my graduate studies when this occurred seemingly overnight. Outside graduate seminars, my colleagues turned toward those few geographers writing on critical approaches to technology, including Rob Kitchin, Stephen Graham, and Sarah Elwood. There was more to the story than was offered within the mainstream of GIScience scholarship, pedagogy, and funding streams. The methods to investigate these changes required a productive confrontation of the epistemology of GIScience—a hard look in the mirror, so to speak. What emerged were hybrid fields that worked to make use of these new infrastructures, technologies, and data to understand sociospatial phenomena while also investigating the implications of this emergent technoscience.

The problem was not necessarily that GIS needed a new definition, to incorporate these emerging areas of inquiry into the social life of GIS and geospatial data, but that the process of defining itself should become an object of inquiry. Who benefits? What becomes internal and external to the development process? Under these pressures, GIScience became positioned as a realm of technoscientific development that ignored technocultural relationships at the risk of becoming irrelevant: *How would we work with our geospatial databases on the iPhone?*

Therefore, rather than identifying what is and is not GIS, I suggest a continuum of technologies, to include desktop software for the collection, storage, analysis, and representation of geographic phenomena and various ubiquitous and pervasive systems that extend these capacities onto the body and the landscape, enabling the governmental ordering of the neighborhood, the city, the state, and the planet. Shifts in the technical capacities of these devices and software also parallel shifts in society: the privatization of publicly funded scholarship, the reorganization of academic disciplines under neoliberal governance, and a general malaise that drapes itself over contemporary public life. The charge of this book is to examine these shifts as they challenge and change our focus, by examining pervasive technologies that are also largely opaque. This is my call to take responsibility for digital mapping in ways that do not deactivate scholarship but make it more resonate.

Resonate, and be realistic. My response to these shifts in the technoscience of digital mapping and the technocultural relationships that permeate these practices is leavened by a sense of the impossibility of fully apprehending or even comprehending the present. Here, I am reminded of Lefebvre's reflections on the Bauhaus and their contributions to conceptions of space. Attempting to understand the important shift in the concept of space that occurred in Dessau, Germany, in the 1920s, he writes:

> The Bauhaus people understood that things could not be created independently of each other in space, whether movable (furniture) or fixed (buildings), without taking into account their interrelationships and their relationship to the whole. . . . Space opened up to perception, to conceptualization, just as it did to practical action.[9]

For Lefebvre, the work of the Bauhaus was to rethink the relationship to site and context, recognizing the specifically *conceptual* work of design—to push thought and action. However, Lefebvre is careful to attempt to understand the implications of this contribution—of highlighting the interrelationality of things in space. He continues, reading against the grain of Bauhaus intentionality:

> When it comes to the question of what the Bauhaus's audacity produced in the long run, one is obliged to answer: the worldwide, homogeneous and monotonous architecture of the state, whether capitalist or socialist.[10]

Indeed, this passage by Lefebvre presents a cautionary tale. Historians may consider our current efforts and our intentions with digital mapping, however emancipatory and progressive, in a different light—and it may be impossible to fully understand now what kinds of futures might unfold within our location-aware society.

Again, there is more to the story. So many stories, in fact. (My own attempts at storytelling are hopefully understood as partial and incomplete—such a confession is always the case.) Perhaps I have felt at times too comfortable within the frame created through the stories told by GIS & Society and critical GIS scholarship. Perhaps a little experimentation with theory and the shape shifting in these fields is urgent: of subjects becoming objects and objects becoming subjects.

New Lines calls attention to these shifts, by drawing on our current preoccupation with the digital map, or the glowing blue dot on the maps of our smartphones that tell us: we are here. Mine is the most basic recognition of an affect—attention and our ability to shift to what we pay attention—as our strongest potential. To lean forward toward a geographic representation is to engage. To be attuned to this simple bodily action is to flatten our discussions of the impact of mapping. Certainly, maps have moved nations, sparked battles, and have connected consumers to products and services. But before lunging at the map and the technical practices that produced it, as a vehicle for colonialism or capitalist expansion, this project begins and ends with this simple moment of engagement. The map is an event.

These events are conditioned by the spaces and times of engagements past. To be clear, they are not determined; however, to engage in the present is to conjure the engagements of the past and of the future. In this sense, engagement is always already historical and geographical. To lean forward, to draw and trace the lines that compose us, is to intervene in these prior drawings and tracings. Gunnar Olsson infects my thinking here:

> Finally: Is this geography? Of course it is! For what is geography, if it is not the drawing and interpretation of lines. The only quality that makes my geography unusual is that it does not limit itself to the study of visible things. Instead it tries to foreshadow a cartography of thought.[11]

I argue for an engaged study of digital mapping as the domain for specific habits of cartographic thought and action. These habits situate a leaning forward to the map, renewing the GIS & Society tradition and the critical GIS agenda as it confronts a society insinuated by ubiquitous and pervasive computing. As such, the study of digital mapping characterizes an approach that seeks to incorporate and move forward on the critique of technologies of geographic representation, by learning about and practicing the technologies with renewed and different approaches.

New Lines is my attempt to explore the conditions and possibilities of the drawn and traced line, to stay with the trouble of the map. Throughout this text, I think of lines as both literal and figurative, both the product of mapping practices and the conditions for everyday life itself. Lines move us. They direct us. They engage. But importantly, we can intervene. We can alter and even erase their imprint. My thinking is saturated by such possibilities of the line—as an expression of force, affect, energy; as direction and connection; as rhizome, composed of many in one; as something one creates, reads, traces, is captured within, and escapes. Far from wanting to settle what a line is or does, I wallow in the messes lines make, to modestly explore both their novelty and their imminence.

INTRODUCTION

But Do You Actually *Do* GIS?

> I took this course expecting technical training in mapping software. The course turned out to be more theory and human geography based.... Our project has immersed me into real world GIS applications, while it may not be technical, it has still been a great experience.
> —STUDENT ENROLLED AT THE UNIVERSITY OF KENTUCKY

> Great teacher, bad course. Drop 95% of the theory, history, influence on culture and research—if you want that, make a separate class.
> —ANOTHER STUDENT ENROLLED IN THE SAME CLASS

Utility. Its expression in the classroom acts as a piece of rhetoric to justify decisions regarding the value of particular curricular paths. Also heard among university administrators—as well as its close relation "relevance"—it works to legitimate decisions about recruitment and staffing in the midst of a series of crises of confidence in higher education. Public scholarship. University engagement and outreach. Indeed, campuses and the scholarship they support are impacted by these expressions. Either bolster support for scholarly activities that fit the current model of value in the academy, or change the conversation. Arguably, many institutions of higher education are attempting the former, reorganizing university structures to better capture resources. There are winners and losers. Witness the collapse of departments within the liberal arts (such as classics and languages), amid the growth of athletic department budgets and expenditures on various other edutainment on college campuses.

These tensions are felt in everyday conversation in the corridors outside lecture theaters. "But do you actually *do* GIS?" Loaded in that question is a series of assumptions about what it means to practice, and I think we can do more to broaden our vision of that practice as both technical and critical. This, of course, necessitates a shift in undergraduate and graduate programming as well as changing the conversation around faculty recruitment. In many ways, private industry is more at ease with these shifts, as organizations that foster geospatial research and development demand a more interdisciplinary vision for their knowledge-workers. To be relevant, to be utilizable, indeed, to be practical, should not have to come at the expense of being contextual, creative, projective, radical, and critical.

Those who ask this question likely expect weighty results on the implementation of this or that mapping technology, or at least a heavier dose of geotechnical jargon. I understand their confusion. I introduce myself as a GIScientist, but one that studies *the use* of GIS. As a result, my work is largely evaluated by my peers in critical GIS and critical technologies studies within Geography, while other GIScientists likely do not include my work within the fold; again, I understand. In the context of blunt comments that question "utility" or "relevance," GIS appears to be on the winning side of these conversations, and yet my hybrid position as a scholar that studies GIS as an object, while teaching courses in the technical practice of GIS, might be perplexing.

But why? I suggest that these conversations are conditioned by the specific history of geospatial technologies within geography departments (specifically in the United States), and I argue that those of us in the discipline need to do better to change the questions that invoke value in the academy to make the strange and perplexing more familiar.[1] To do so requires our best sciences, arts, and humanities, a widening of interdisciplinary opportunities, and an invigoration of the liberal arts in a new generation of engaged scholarship.

Positioning

The increasing availability of and innovations in internet-based digital mapping tools have brought about rapid changes in mapping practices. Alongside this popularization of mapping has been a largely silent academy as to what these developments mean for cartography, GIScience, geography, and spatial thought more broadly. Meanwhile, the arts, humanities, design, and social sciences, including critical human geography, have marked their interest in the use of geospatial technologies, with the emergence of map art exhibitions, renewed academic subfields like the digital and spatial humanities, as well as calls for new collaborations between the critical social sciences and the GISciences.

Therefore, this book takes as its central organizing question: what are the new conceptual and theoretical challenges for the critical engagement of geographic information systems? Since the late 1990s, critical GIS has referred to an area of research positioned at the intersection of critical geography and geographic information science, drawing together technical capabilities for geographic representation and analysis with the critical capacities of social theory and, recently, more-than-human geographies, the digital humanities, and digital geographies.[2] Critical GIS scholarship is particularly influenced by the work of participatory action researchers; the histories of cartography and geographic information technologies; and the inclusion of radical, local, everyday knowledges. It grew through feminist geography's and feminist geographers' insistence on the conditions of knowledge production and representation and the promotion of alternative methods and epistemologies. It inherits a focused attention to the social implications of geospatial technologies from the GIS & Society tradition while being cognizant of the technical debates and intricacies of GIScience.

While some may feel that critical GIS is a project that has evolved beyond what is indicated by its naming, I continue to suggest the political importance of this intervention signaled by its name.[3] Critical GIS is a field centrally concerned with the proliferation of digital mapping opportunities, both as phenomena worthy of careful study and as the development of capacities for enacting critical, radical, or simply alternative knowledge-making endeavors. This book narrates a critical GIS that is directed at the tensions which have persisted in the GIS & Society tradition, a tacking between the use of geographic technologies for projects of representation and the representation of geographic technologies themselves. This particular tension motivates a renewed agenda for the field, of how to advance a spatial project that opens up the technology to "new" practices while providing the critical and conceptual tools necessary to situate its responsive and responsible use.

The distinct practices of critique and application are not necessarily oppositional activities but are often represented as such. Indeed, one historicization of critical GIS might point to the emergence of critical GIS as a resolution to the deconstructive critiques of GIS in the first half of the 1990s. In other words, according to this narration, critical GIS scholars sought to take action in responding to the critiques of GIS by *doing* GIS differently. Of course, some critiques were found to actually deactivate

GIS itself; for instance, calls to recognize the use of GIS to enact military violence and imperialism might be read ultimately as a call to dismantle the technology, or in the least, disrupt the complicities of the discipline of Geography to yet another project of colonial imposition.[4] Still others found an opportunity in developing the science of geographic information systems, crafting a subfield that could capture the scholarly work of both developing the technology (its interface, programming structures, and tools) and applying the technology to new research questions of the social and natural sciences.

What follows is not merely a retracing of the literature that formed critical GIS nor is it simply a survey of the contemporary ways in which a critical GIS is being practiced. Rather, this project advances a "history of the present" with regard to critical GIS, in order to clarify its agenda, to take a position on the evolving implications of digital map use in everyday life, and to argue for a renewal of critical GIS despite its distractions and detractors. To conduct such a project, I forward two presuppositions. First, as systems of geographic information, GIS encompasses a wide range of technologies for the representation of geographic phenomena. This includes more traditional understandings of GIS as a desktop software application such as Esri's ArcGIS as well as a range of distributed, mobile, cloud-based, and so-called map 2.0 technologies, such as Google Earth/Maps, satellite navigation systems, smartphone mapping applications, location-based services, and so on. Second, this broadening of what is meant by GIS then requires a renewed intellectual apparatus to both theorize and practice these technologies of geographic representation.[5] In other words, this project argues that significant shifts have happened around technologies of geographic representation, beyond new software versions and new hardware capacities and capabilities. Rather, these shifts are described as *technocultural*.[6] As such they demand a renewed critical GIS agenda that shapes practice and research, as well as teaching (as the application of critical GIS finds its way into GIS classrooms). What do these new possibilities for interaction with geospatial technologies mean for engagement and outreach, for participation and collaboration, for alternative spatial knowledge creation, and for responsible mapping?

To address this question, I suggest a four-part critical GIS agenda. First, critical GIS is an interrogation of the social, political, and economic conditions that enable our contemporary notions of geospatial technologies (both as histories of GIS development and the fashioning of digital spatial thought). Second, critical GIS is an evaluation of the concepts that inform or are assumed by contemporary GIS practices (including representation, collaboration and participation, standards and interoperability, data and information, privacy, and so on). Third, critical GIS examines the multiple ways in which GIS is enrolled (including as location-based services, as geointelligence, as crowd-sourced citizen science, as smart city, and so on). Fourth, critical GIS is productive of postures and practices, alternative pedagogies, community-based mapping, qualitative GIS and geovisualizations, and everyday mappings. In renewing the critical GIS agenda, I situate the emergence of geospatial technologies alongside the fashioning of critical GIS, to examine the ways in which critical GIS is currently practiced, to present the persistent conceptual problematics of critical GIS as expanded toward an engaged, digital mapping studies, and in doing so, to connect critical GIS to the emerging literature of the digital humanities, technology studies in geography, STS and technoculture, the application of critical thought, the (new) materialization of cultural theory, posthumanisms, digital geographies, and so on. To engage in a renewal of critical GIS research is to do more than shift attention within GIScience, or even within geography. Instead, as this project suggests, there is an opportunity to reach beyond the disciplinary interests of geography and GIScience and curate a collective focus, a leaning forward, toward the mattering of planetary survival, itself.

Being asked the question "But do you actually *do* GIS?" unearths a series of broader concerns, however. Its posturing is indicative of a growing, nagging skepticism of the project of higher education that has particular impact on the composition of academic departments and the kinds of scholarship that are furthered. To fashion a response to this question has meant reconsidering what it means to be "public" as well as a "science" in a democracy. In this project, I reevaluate our disciplinary stance, to challenge the definitional boundaries of geographic information systems and to reestablish the significance of *studying* versus *doing*. Follow the lines, while reading between them.

Lines and Limits

At first glance, cartographers and map enthusiasts may find little to work with in Gunnar Olsson's *Lines of Power, Limits of Language*.[7] The book unfolds as a philosophy, bordering on theology, inasmuch as it develops conceptual thought while also investigating belief. And while his research program is differently aligned than more clearly bounded projects of critical mapping, I feel his discursive moment in a concluding passage:

> And now, toward the end of the beginning, it should be clear what fascinates me. . . . Perhaps I enter this social space of silence by living

in the world as I found it. A world where the unconscious is structured as a language, a world where power is structured as a knowledge, a world where lines are taken to their limits.[8]

This sentiment is about recognizing the intense and fantastic responsibility that comes of being a geographer, of conducting the course of lines past, present, and future. Olsson elevates the cartographic as central to organizations of knowledge, examining symbols and signs as systems of meaning.[9] The relations between word and world and the curatorial systems that map these relations are brought to the surface by discussions of the line. Extending Olsson, "is" and "=" (as two such lines) enact much trickery in the GISciences, signaled, for instance, by the debate between Agnieszka Leszczynski and Jeremy Crampton on the role of poststructural critique to understandings of the materiality of GIS and mapping practices.[10]

The problem instead, as Olsson writes, is how to "recognize something when I see it again," when "the phenomenon under discussion changed as the story proceeded."[11] The scurry of current practices marked under the sign of *GIS* are such phenomena; among them, changes for the line must also mean changes for techniques of recognition and engagement. In other words, the conditions for digital mapping and spatial representation have indeed changed, and scholarship about such technologies and their conditions will need to take different forms. This insistence scratches up against my background of teaching critical GIS and facilitating university–community partnerships in GIS classrooms. While this work has to date emphasized the sociopolitical dimensions of spatial data and technological practices, I am increasingly interested in how to foster and sustain a critical technology perspective *within* GIScience—instead of conducting this work in the shadows created by the field. The challenge then is to allow the thickening of the line represented by GIScience practice: first, by recognizing that there *are* multiple implications of GIScience, and, second, by deepening our attention to the conditions, status, and effects of such work.

The last century of mapping has brought about incredible opportunities, with a frenzy of disciplinary subfields that travel under the banner of mapping, and yet extend into many modes of thought, inquiry, representation, and habit. Across many of these developments has been a particular theory of action—which might also be called practice or method—wherein a difference is made and the world is intervened.[12] From the drawing of a line to connect distant nodes to the analytical and interpretative practices that trace drawn lines, the spaces and places of our discursive and material realities are formed and informed. I understand drawing and tracing as fundamentally distinct, yet complementary, theory-actions, that the drawing and tracing of a line is both territorializing and deterritorializing, both stratification and destratification, to invoke Deleuze and Guattari.[13]

As forces or affects, these theory-actions that compel both the drawing and tracing of a line are always beyond a simple symbolization of the real. "It is tracings that must be put on the map, not the opposite."[14] Indeed, there are underlying interests of drawn and traced lines, but there is more potential in these actions. For Deleuze and Guattari, lines are what make up their rhizomology. Here, they make distinctions with tree structures, as a pathway beyond dualisms that had infected much social theory: neither subject nor object, faith nor reason, signifier nor signified. Instead: *and*. Rhizomes as conjunctions and compositions of lines are "directions in motion," conjuring maps (and here, I am *also* thinking actual map artifacts) "that must be produced, constructed . . . always detachable, connectable, reversible, modifiable."[15] These maps become rhizomatic, in that they are composed of lines that stratify, in the least, and that at their greatest potential create a "line of flight or deterritorialization . . . after which the multiplicity undergoes metamorphosis, [and] changes in nature."[16] That maps stratify *and* create the potentialities for resistance has long been understood within critical cartography.[17] However, how best to connect these as theory-actions? How might a potential be recognized and released? There are those who draw and those who trace; each practice has a politics, and their attendant black boxes confound the novice. Representations of this sort are thus always already more than representational.

The line, drawn by us, in turn, draws us in. The subtleties of a drawn line and the affective force of its presence or absence is integrally part of such a theory-action that stretches across and assembles the various forms of mapping practice. In other words, I believe Olsson is accurate in directing our attention to perhaps the most banal of a GIScientist's endeavors: "For what is geography, if it is not the drawing and interpretation of lines?"[18] However, in the midst of an increasing proliferation of digital tools, techniques, and literacies of mapping, I read Deleuze and Guattari to find the significance in a particular responsive and responsible action: the enacting of a social history of GIS. Put simply, this is a call for tracers who draw and drawers who trace—to put tracings back on the map.[19] It is a recognition of the urgency of such hybrid positions, of "the promises of monsters."[20] Haraway considers such a "mapping exercise and travelogue" in order to:

write theory, i.e., to produce a patterned vision of how to move and what to fear in the topography of an impossible but all-too-real present, in order to find an absent, but perhaps possible, other present.[21]

To do so is to return to the potential of the GIS & Society agenda and the Friday Harbor meetings that catalyzed it, to confront the opacity and rigidity of our present GIScience (as well as its short-term, selective memory).[22]

A social history of GIS aspires to be resolutely situated in the disciplinary histories that encircle GIScience, to recognize that the GISciences are not hermetically sealed, that they emerge from within specific traditions of innovation and investment. The significance of this social history is a remnant of the original GIS & Society agenda—lost in the fog of GIScience rhetoric. Indeed, John Pickles, in reflecting on those days more than twenty years ago that gave rise to *Ground Truth*,[23] discusses the atmosphere of geography departments:

> GIS students were rarely introduced to the prevalent debates about philosophies of science, social theory, and cultural studies.... In parallel, the technical possibilities for larger data-sharing and analysis were not taken up by most Marxist, feminist, and humanistic geographers.[24]

Still more lines drawn, divisions articulated and reinforced. This reflection could be as easily made about our current moment as it was made about 1993. The point is that *this* social history perpetuates—that more hybrid positions like Marxist GIS or humanities GIS were not thought. How could they possibly be acted? Disciplinary events like Friday Harbor or *Ground Truth* or Initiative 19 (I-19) of the National Center for Geographic Information and Analysis were meant to inspire such hybrids, such monsters, such troubles?[25] What promises did these events hold?

As a student at the University of Washington in the decade-plus following these interventions, I considered moments like I-19, on "the social implications of how people, space, and environment are represented in GIS" as endeavors to figure out how to "work together."[26] And there have been synergistic outcomes of this ethic of transepistemological engagement, and yet a social history largely resorts to the well-etched lines that separate and subtract. The point of enacting another social history is to allow other restratifications, not to find *the* binding lines but to allow the field to become more plastic, indeed to become. New lines, not novel ones, but imminent, coming, projecting.

INTRODUCTION 9

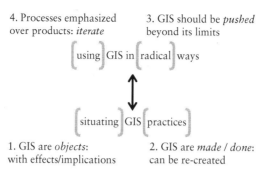

FIGURE 1. Four provocations in the tension between using GIS in radical ways and situating GIS practices.

One of these restratifications moves under the name of critical GIS, which I consider a mode of inquiry that is a tacking back and forth between technical practice and critical practice. I have previously called this a technopositionality, between using digital spatial technologies in radical ways and relentlessly situating those same technological practices.[27] This tension between the use of GIS and the study of the use of GIS enables four further provocations that I believe enact this ethic of "working together." See the diagram in Figure 1. First, GIS are objects; it is assembled as an institution. As software, it therefore has effects and implications, and one can trace its operation and constitutive role in society. Second, GIS are made but not made up; it is done, enacted, produced, constructed. As such, it can be re-created and made differently. Third, GIS should be pushed beyond its limits, to constitute new, alternative technologies and radical implementations. However, critical GIS requires practice. Fourth, it assumes iteration. Processes are emphasized over products. These products are simply souvenirs of mapping journeys; they are the result of momentary assemblages.

These four provocations have not been equally taken up. Indeed, the first requires a particular historical method and conceptual footing that has, at times, loosened the connective threads that bound the GIS & Society tradition. Furthermore, the sedimentary processes by which critical GIS has become "a thing" (its stratification) has laid ground for some interesting, sometimes parallel, developments within and without our discipline, including feminist, queer, qualitative, and historical GIS, as well as GIS art, nonrepresentational GIS, and the spatial digital humanities. There have been renewed alignments with participatory action mapping as well as an emerging critical physical geography.[28] New work in critical geoweb studies has reinvested a generation of scholars in the geographies of ICTs and digital culture, while the situating of neogeography has led to pedagogies in critical GIS and mapping.[29] These areas share either important antecedents or an approach, a conceptual practice of mixing or a recognition of the deeply transitive and translational properties of engaged technical and technoscientific work.

Thus, critical GIS is always a doing and an undoing. This dual characteristic, this productive schizophrenia, I believe, makes it a unique technopositional stance. As a result, critical geographers are drawn toward GIScience, as a way to analyze issues related to social and environmental justice, recognizing the need for "strong" geographic representations to articulate both global uneven development as well as the injustices of everyday life for some. Relatedly, the digital humanities, in their turn toward the geohumanities, use locational indicators to understand novel structure, character development, and the relationships among traditional artifacts of humanistic inquiry. On the one hand, this is still about the necessity of "working together."

However, another "gorilla" has entered the room—perhaps more forceful and broader in disciplinary desires than the mandates of science in the late 1990s debates around GIS. This requires, I argue, the reprioritization of a social history of GIS within the critical GIS agenda, an underlining of my first provocation. The infographic in Figure 2 highlights big data market forecasts, global mentions of big data on Google Trends, and insights for ecosystems and businesses, not just questions, but the *right* questions. We need to tame big data. Indeed, it is hard to deny that the GISciences (alongside seemingly every corner of the academy) are being rearticulated through an emphasis on big data and the opportunities associated with user-generated Internet content, through funding for cyberinfrastructural systems, computational methods for securing the homeland, and the gaining of geopolitical high ground through geointelligence services—not to mention the range of industries that feverishly inspire GIScientists to better accommodate digital economies based on attention, with real-time geodemographic, location-based support services.[30] Perhaps, as Nadine Schuurman and Mei-Po Kwan suggested in 2004 and as David O'Sullivan asks more directly in 2006, this move to rebrand geotechnical scholarship and practice as GI*Science* has actually disconnected GIS from the sociopolitical and economic situations of its emergence.[31] Proponents of big data science articulate rigid lines that serve to render the GISciences more opaque, more available to the opportunities within our digital economies.

Here, my thinking is animated by these lines, as moments of stratification and reterritorialization. Figure 3 crudely traces the lines that draw in

FIGURE 2. Wikibon created an infographic on the topic of big data and its relevance for business and advertising, including current uses, a market forecast, and global mentions in Google trends. "Taming Big Data: Big Data Includes Data Sets Whose Size and Type Make Them Impractical to Process and Analyze with Traditional Database Technologies." Reprinted courtesy of Jeff Kelly and David Butler, http://wikibon.org.

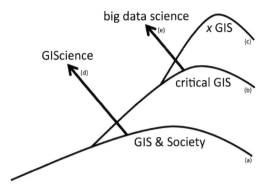

FIGURE 3. The discursive terrain of GIS is marked by stratifications and destratifications.

GIS with departures and responses. If GIS & Society laid the foundations for critical GIS, from which a whole host of x-GIS subfields and activities have emerged (such as feminist, queer, and qualitative GIS), then how we view the rebranding of GIScience in the late 1990s[32] is important for our current encounters with big data science decades later.[33] If these lines are even partially true, how might we respond? Was Friday Harbor in 1993 (and again in 2014) just one of many opportunities to change direction, to gather energies, to destratify? Have we lost that affect, that force, to work together in drawing and tracing lines?

While there may be multiple ways in which to imagine new lines, new practices of engagement with and through GIScience, I tend to approach this by asking the following question: How do we *enact* a social history of GIS? Which is to say, how might the social history of GIS enable new practices and engagements that mobilize GIScience, that cause mapping to move? A social history of GIS tacks toward the first provocation, toward the tracing of operations and effects of GIS. However, to *enact* a social history of GIS is to allow this social history to actually impinge on GIScience practice, to thicken its remit, to put the tracings back on the map. What do these new lines look like? Perhaps they are pedagogical, taking risks in our classrooms to interrogate alongside our community-based partners—allowing a kind of vulnerability, to change course, to map the yet unmapped. And doing so, by following mapping industries, perhaps through participation and observation, or working to develop alternative techniques that inspire a renewed responsibility to the various geographies we inhabit. Of course, there are many other possible lines that we should explore.

The suggestion I am attempting to make is that perhaps geographers have enabled a particular and limiting narrative that surrounds the practice of GIS—as either science or critical—and have produced the range, the terrain, of possible engagement. As Deleuze and Guattari write, "A rupture is made, a line of flight is traced, yet there is always the risk of finding along it organizations that restratify everything."[34] The ruptures generated by GIScience and big data science are moments of flight to create new potentiality in a field that was too certain of itself. However, these moments risk restratification—reestablishing precisely the kinds of insider-outsider contexts that vexed the discipline in the 1990s. As Deleuze and Guattari suggest, such new lines may eventually become hardened in the formation of strata, but the point is to pursue these lines of flight, to take these risks, and take them as often as permitted. As an advocate of the technopositional in GIScience, I argue that our field may be limited only by our ability to produce these new lines. The promise of monsters—of drawers who trace and tracers who draw—is that they are always rhizomatic, always productive of continuous and resistive destratification. But this is a heavy responsibility, with nontrivial implications for pedagogy and practice. Perhaps this is why Pickles's observation on departments in 1993 is so apropos—students of GIS are still rarely given opportunities to consider social theory while social theorists feign an allergy to technical practices. Deep strata, indeed.

Perhaps these new lines will return us to the map: the map as an event, a kind of aporia, a difficulty, a perplexity. In this sense, on the one hand, the map is recognized as an artifact that captures and assembles a vision such that other representations, other imaged and imagined landscapes, might be made invisible. However, on the other hand, and with great pause, the map artifact draws one in, causing one to actually lean in and trace the contours of place. You are here, and the possibility of a collective, of a "we are here" is not unimaginable. A responsive and responsible cartography constitutes a blurring of the individual and the collective, to resist the simple distractions created by the glowing blue orbs in the interfaces of our mobile devices, by rendering more transparent the assemblages that make its pulsations possible. Therefore, the "we" of this imagination is resolutely more than human. How I think the enacting of such a social history of GIS proceeds through this resolution, pushing our greatest technoscientific achievements to inspire collective engagements, to bring about worlds that make a difference, where that difference is always contingent, in "a world where lines are taken to their limits."[35]

When Rats Move

The new lines we draw and that draw us in are gritty, and I understand them as qualities of rhizomes, as lines that thicken as "when rats move by sliding over and under one another."[36] At this point, it has hopefully become more obvious that I am thinking of lines not only as metaphors for directionality or borders and boundaries nor only as literal, drafted lines on a map manuscript. Rather, the lines we draw and the lines that draw us in are of differing qualities. They are continuities. Deleuze writes:

> There is not simply the opposition of earth and water, of the one and the many; there is the transition of the one into the other, and the sudden upsurge of the other out of the one.[37]

To conceptualize these lines, thick and thin, drawn and traced, as merely transitions between qualities is to sidestep, at least temporarily, the various binding crises of representation (objectivity/subjectivity, perception and partial perspective, science and belief, virtual and the real, and so on). In other words, it is not enough to trace the map (as perhaps historical cartography), or study the implications of the map (as perhaps critical GIS), but rather to map the trace, to intervene with new lines, not as representations but as vibrations, disruptions, forces, interventions.[38] To engage in such a project, I suggest that critical mapping must grapple with a fundamental question of practice, to challenge our disciplinary muscle memory that comes to expect a certain kind of mapping software (ArcGIS) and a certain kind of critique ("killing machine"[39]). Indeed, despite the evolutions of both critical theory and technologies, the discourses of a situated technological practice are incredibly anemic. For instance, take Harley's oft-paraphrased statement, "cartography is seldom what cartographers say it is,"[40] where I began in the preface. I understand Harley's project as a kind of provocative excavation of ideology in mapping. While not a theorist, Harley recognized that just below the surface, the practice of cartography was both more and less that how it was being studied and enacted.

In this sense, a Harleian tradition magnifies the cartographic tradition in order to inspect and interrogate the assumptions and the implications of specific practices and representations, to identify the ideological resonance and tensions of maps, mapmakers, and mapmaking.[41] Recognizing that I am perhaps overly glossing of a Harleian tradition, I wonder about a different reaction—a different habit and the training of different muscle memories—to cartographic practice.

I explore the implications of this statement: that cartography *is* as cartographers *are*. Different positions, different perspectives, different attunements. I believe that Arthur Robinson, the leader of mid- to late twentieth-century thinking in cartographic pedagogy, understood the importance of the cartographer's craft and how to use scientific rigor to refine and innovate that craft. His work suggests an importance of experimentation with visual strategy, in order to develop a register of best practices regarding geographic representation. In *The Look of Maps*, one part of his dissertation from 1952, he writes of map design:

> The one aspect of map structure which seems to have definite possibilities for objective visual evaluation is the projectional. Heretofore, almost the entire literature on projections has been concerned with the mathematical phases. Only recently has the visual problem become significant in the selection of projections.[42]

Despite the sense that contemporary cartography has moved several leaps beyond the Robinsonian moment, I suggest that neo-Robinsonians are very much interested in these convictions—that while our technologies of cartography have shifted and accelerated, the need to be deliberate and objective in design decisions remains. I want to explore "objective visual evaluation" as an alternative perspective on the proliferation of mapping technologies and geospatial data and the permutations of the mapping industry amid shifts in higher education. How might the projection, as the hallmark of an engaged mapping, function? How might map users understand our curved planet and the spatial relationships born on the ground? What are the unique technopositionalities that are enabled? What are the implications for technical and critical practice?

In taking up these questions, I want to not so quickly dismiss the "objective visual evaluation" that motivates much contemporary scholarship on cartographic behavior—so-called map use research. I suspend my disbelief in such an objective gaze in order to learn how that evaluation functioned and formed disciplinary strata. Indeed, there are different affective modes that are often elided by map use research. My curiosity about such fractures begins with a moment I first made note of over decade ago while engaged in a project examining the ways in which community organizations made use of mobile devices to project alternative futures for their neighborhoods.[43] In Figure 4, taken from training materials used by the Fund for the City of New York in their quality-of-life indicator program in the early 2000s, we can see a group of citizens convened to be trained in the use of a mobile device for the mapping of assets and deficits in their neighborhoods. While this earlier research sought to understand the interlocking ways in which technologies and neoliberal governance served to recast who or what bodies belonged on the sidewalk, I return to this image to illustrate a most basic mapping relationship.

FIGURE 4. From the training materials of the Fund for the City of New York in their neighborhood indicator project. Reproduced courtesy of Sustainable Seattle.

This photo represents a familiar scene for me, over several years of community-based mapping work. At specific moments, individuals feel compelled, consciously or not, to lean forward in their seats, toward the image before them. A few rows back at the left of the photo, a woman tilts toward the front of the room, signaling, perhaps, an active listening, a visible engagement within a family of nonverbals like the nod or a furrowed brow. These affective moments (might they even be called mapping interactions?) are not so easily understood or objectively measured—despite the several decades of behavioral research on map use. These moments resist taxonomy. They make mischief of "objective visual evaluation." They are only ever transitions, repetitions, forces of one object against another.

Rebecca Krinke's project *Unseen/Seen: The Mapping of Joy and Pain* at the University of Minnesota highlights an analog and deeply tangible process of engaging in the emotions of representation.[44] Participants added color to a blank map canvas to represent their feelings of joy and pain in the Twin Cities, seen in Figure 5. They lean forward, discuss, debate, contest, and celebrate. These moments are prized and elaborated in psychogeographic projects, to include map walks, field papers, and other dérives or drifts.[45] Similarly, Figure 6 is a photo of a participatory mapping

FIGURE 5. Rebecca Krinke's mapping project *Unseen/Seen: The Mapping of Joy and Pain*. Photograph by Rebecca Krinke, Mears Park, Saint Paul, Minnesota, 2010. Reproduced courtesy of Rebecca Krinke.

project I facilitated in Edinburgh, Scotland, in 2012. After going on short, timed, unguided walks, participants traced a projected map with their fingertips and placed their thoughts, feelings, and projections directly onto the map surface. These traces became new lines, new encounters, new interrogations, more than simply new representations. These participants put their traces back on the map.

Here, now. These new lines thicken the moment of representation. They both assemble our thoughts and actions and aid their dispersal. The point is to attend to these thickenings, as when rats move, to recognize that the ineffable qualities of engagement with the map will always resist "objective visual evaluation," and yet will of course also always make such evaluative measures possible. Deleuze and Guattari write,

> Because one never knows in advance how a line will turn, politics is an experimental activity. Make the line break through, says the accountant: but that's just it, the line can break through *anywhere*.[46]

We turn toward the map in front of us, because it compels us. Maps move. They organize bodies. They channel and focus. They damage and recover.

FIGURE 6. A psychogeographic project in Edinburgh, Scotland.

At times, we experience and craft them as solids. In other moments, they are plastic and fleeting. The potential of mapping lies in the slippage between solids and plastics. The decade-plus of research in public participation GIS has taught me that. Legitimacy is something you craft, not inherit.

And yet the space of potential radical engagement through cartography has become crowded. We are not just in the company of a few critical and counter-cartographers or critical and participatory GISers. Instead, as the rise of new mapping players in the geospatial industry and new applications of critical human geography and participatory mapping indicates, this space of potential in mapping is not entirely (nor was it ever perhaps) ours for the making. The crafting of legitimacy, the making of new lines in this space is ever more complicated. Far from allowing these developments to simply wash over us and push us aside, I still believe and insist that the academy has a place in this debate.

However, those of us in the academy have become distracted, as well. In the context of blunt comments that invoke "utility" or "relevance," GIS appears to be on the winning side of these conversations, although perhaps not critical or counter in approach. Inasmuch as we need to draw attention to the various industrial tensions and pressures on mapping, we need to shift the registers for what counts as important, relevant pedagogy in our departments. We need to change this conversation, without abandoning the incredible innovations in and outside the classroom computer lab. These tensions around GIS regarding its place in the university is not only typical of the neuroses of the discipline but is also illustrative of the wider and urgent problematic of the role of higher education.

Utility and Relevance

The tools collapsed into what is called GIS have been taken up by the humanities and social sciences, as well as by the design disciplines of architecture, landscape architecture, and urban planning. As a result, the scope of these tools has been diversified well outside the earth sciences. Despite the relatively brief pushback experienced in some geography departments in the 1990s,[47] course offerings and faculty positions in the area of GIScience have stabilized, if not greatly expanded. This solidification of GIScience within universities parallels neoliberalizing struggles within higher education and underlines the unevenness of institutional investment in university subjects and the impact of the discourses of utility and relevance.[48]

Consider the advent of massively open and online courses (MOOCs) and the recent incursion of capital into the production and facilitation of these classrooms. Anthony Robinson, geography faculty at Penn State University, launched the MOOC "Maps and the Geospatial Revolution," beginning with around thirty thousand students on July 17, 2013. In a *Wired* interview, he discusses the motivation for creating the course:

> When I meet someone on a plane and tell them I'm a geographer, they're like "What?" They don't even realize that's a thing. Something like a MOOC, that's free and has a high profile, might get more people interested in what we do.[49]

The anxieties of discussing one's academic profession to a lay audience is not necessarily unique to those within the discipline of geography, but geographers for the better part of two decades have leaned on the technocultural popularity of geospatial technologies in providing signposts for a curious (even skeptical) public as to the value of a geographic education. The question of "what we do" is an already loaded one—assuming that a "doing" (versus an "undoing" or a "do nothing") is the greatest social good. But even further, the saturated language associated with geospatial education (consider the scope of education consultants at Esri) mimes sentiments that surrounded the advocacy of GIS within departments of geography in the early 1990s: GIS will save the discipline from itself.[50]

However, this sentiment has extended beyond the confines of geography departments and has been further hailed as a way to resolve the crises of the humanities and the social sciences in the wake of pressures to "do something" and "be public," to be both utilizable and relevant. A basic online search of *relevance* within the websites of *Insider Higher Ed* or *The Chronicle of Higher Education* will reveal a persistent conversation that parallels a discussion of the sustainability of departments as well as the viability of specific disciplines. However, in a limited way, these pressures come preformed to higher education fashioned in the mechanisms that prepare students for the university.

For instance, even the standardized examinations for college admissions move toward a line of questioning that underscores the shifting registers of collegiate knowledge that is both expected and fostered:

> Every year, the SAT reduces more than a few teenage test-takers to tears. But few questions on the so-called Big Test appear to have provoked more anxious chatter—at least in this era of texting and online comment streams and discussion threads—than an essay prompt in some versions of the SAT administered last Saturday in which students were asked to opine on reality television.[51]

That students were asked to react to and discuss reality television is indicative of more than a targeting of the audience of such exams—to assess writing and thinking skills using topics of current and general prominence. Instead, this passage indicates a much more sinister development, central to the fashioning of what society considers general knowledge. To ask students to develop an opinion on drone warfare, foreign policy, or global economic crises would be to complicate the evaluation of basic thinking and writing skills. Instead, American youth are said to have "kept up" with the Kardashians, and therefore can be expected to have thoughts and even arguments to wage in response. This is most curious.

In this context—of a simultaneous winnowing of the registers of knowledge *and* a rescaling of the metrics through which we recognize intellectual development—we can place the attenuation of a geographic attention while the technological conditions of the map seemingly flourish. How might the map both contribute to and disrupt the attenuation of a geographic attention? What about being relevant and utilizable both champions the map and leads to its undoing?

Within higher education, budgetary models reward units that fill classrooms: the butts-in-seats model. The implications for this are felt most strongly perhaps within the languages and the humanities faculties, where the development of critical thinking and language skills require a more favorable ratio of faculty to students. Even still, the humanities are encouraged to modularize their lectures and classroom activities, incentivizing a move that attempts to economize faculty labor and, often simultaneously, distill the content of a lecture into action and practice, aligning the work of critical thinking with the development of technical facility.

Pressure to reshape curricula toward utility and relevance occurs most incrementally in the workshopping of course syllabi—often through teaching and learning centers on campus—to develop active verb learning objectives. To argue that students will learn, understand, or even know is, according to the regimes of syllabi authorship, too conventional to attract students as well as too abstract to be assessed. Sure, students will learn, perhaps understand, and even know, but more savvy syllabi will foster action and practice. Students will *do,* not *just know.*

At a more intermediate level, funding is targeted at faculty to redesign their courses—to invert the classroom, to build in project-based learning, to partner with industry as well as community organizations, to coteach multidisciplinary courses—in order to prioritize learning that is actionable, to alleviate the concerns of parents and the student-customer as to the efficacy of a liberal arts education. Teachers will *demonstrate,* not *just evaluate.*

The GISciences have benefited from these pressures as instructors and researchers are tacked on to curricular innovations. Within this emerging utility regime, the GISciences are made multiple through the various partnerships and alignments that are possible across the humanities and social and natural sciences, as well as in professional fields of public health, medicine, engineering, architecture, and design. Coursework enhanced with GIS provides students with an opportunity to *do* in the classroom, shifting the underlying relationships between students, faculty, and a collegiate education.

The advent of public scholarship, particularly within the humanities and social sciences, frames another response to these problematics. In a debate with Ash Amin and Nigel Thrift on the direction of the Left in academic geography, Neil Smith takes aim (again[52]) at GIScience:

> We still live today with the bountiful results of the broad social theory revolution in geography and the discipline is a far better place for it. The multiplicity of social theoretical perspectives makes it an enviable domicile compared with the doctrinaire narrowness of economics, say, or political science. Gone since the postwar era is the withering definitional retort: "but is it geography?" And yet a significant backlash has already set in. Some of it rides on the back of Geographical

Information Sciences (GISci), reasserts the power of a narrow scientific positivism, and reframes the discipline as a spatial science in the service of technocratic power.[53]

Smith asserts a return of the debates that surrounded the incursion of GIS into departments in the late 1980s and 1990s.[54] In the context of this debate, however, his statements draw these concerns into a broader set of questions about the role of academics in speeding along the processes of neoliberalism within universities, and thereby enabling a dismantling of the radical agenda of a prior generation of Marxist geographers. GIScience signals this tension between the recasting of geography as a spatial science and geography as one vehicle for public scholarship. An alternative trajectory needs drawn.

Fractured Lines

I suggest an occupation of the space of the line. To find potential tangency between the mapping sciences and the affective resonance of drawn lines is to create something anew. Our way finder for these points of tangency should measure vibrations. How does a body lean forward toward the line? How might that leaning signal interest, investment in, conviction about a subject? This gesture—to lean forward—should be the hallmark of our best mapmaking practices. To recover the plasticity as well as the strategic solidity of the map (its destratification amid stratifying forces), I discuss five fractures or five potential lines of flight that consolidate as well as radiate, that stay with the trouble. Each occurs at different speeds and produces differing volumes in the crevices they create: (1) criticality, (2) digitality, (3) movement, (4) attention, and (5) quantification. There are undoubtedly others; however, these feel most urgent, most troubling.

The first fracture, *criticality*, is a reflection on the stories those of us interested in a distinctly critical mapping tell ourselves about the origins of our subfield. By expanding these origin stories, I attempt to connect critical GIS with antecedents found not only in the rise of a critical human geographic project in the 1980s but also in the more modest response to the quantification of geographic study amid the rush toward computational methods in the 1960s. We have been both *here* and *now* before. The second fracture, *digitality*, represents a project that attempts to recover a particular story about the development of computer mapping usually lost in the wake of narrative about the closing of the geography department at Harvard. Here, I excavate materials associated with Howard Fisher, the architect hired in 1964 by the Graduate School of Design (GSD) to found the

Laboratory for Computer Graphics, which enjoyed a speedy and voluminous impact for a little over a decade: much more than a flash in the pan.

The next three fractures attempt to diagram the current conditions for the digital map, and thereby examine the prospect for leaning forward. *Movement*, the third fracture, travels back and forth between recent developments in design traditions and late 1960s work in animated cartography by Waldo Tobler and late 1970s work of Geoff Dutton, an advisee of Bill Warntz. Here, I invoke movement in both a physical and affective sense: to cause to move, as one object acts on another, and to be moved, as a bodied response. Relatedly, the fourth fracture, *attention*, explores the mapping processes by which attention is controlled and dominated by the cultural industry. The possibility for leaning forward hinges on the capacity for an engaged mapping process. Research with community-based organizations highlights the tensions associated with maintaining visibility, and thereby viability, in a crowded attention marketplace.

The final fracture, *quantification*, documents the exemplary and contemporary ways in which geospatial technologies have been extended toward industry, government and military, and public mapping projects, including location-based services, geointelligence, and social media projects. I take up the rise of personal activity monitors and smart urbanism, by drawing continuities between these developments and the affectations they promote. A new political economy of data, representation, and management permeates the regulatory regimes of the body, city, and nation, and digital technology corporations stand ready to concentrate power and profit. As it did generations earlier, the cartographic acts as the vehicle for this quantification.

I aim to document this fracturing in the politics of the map—from the map as an object for political transformation to the map as an expression, to mappings as the always undone and undoing project of geography. This is the study of the shifting registers of engagement with computer-aided cartography, where a renewed model of interactivity with maps as media drives emerging technologies. To reclaim the map, through such a modest appreciation of its potential, requires a reworking of the tensions that pull at the seams of our discipline—not to "put Humpty Dumpty back together again," as Stan Openshaw quipped two decades ago, but to alter the target of critique, to redraw the lines of continuity that might create anew the energies that inspire a curious public.

This will require more of us—both a greater involvement of geographers and nongeographers and more out-of-the-box thought and action within our ranks. That a single point does not form a line summarizes this requirement. It serves as a kind of manifesto for a renewal of critical mapping practice, recognizing that this work cannot reside only within the single point of an individual, a discipline, or a subfield. Instead this work must—under these new mapping conditions—be part of an expanding constellation, where connecting the dots with new lines is the making plastic of our most prized mapmaking habits.

We need new lines.

CHAPTER ONE

Criticality

The Urgency of Drawing and Tracing

> GIS is not fixed and given but constantly remade through the politics of its use, critical histories of it, and the interrogation of concepts that underlay its design, data definition, collection, and analysis. In other words, futures of GIS are contested and openings exist for new meanings, uses, and effects.
>
> — MARIANNA PAVLOVSKAYA, "THEORIZING WITH GIS"

To draw a line is to make a difference. A line can create left from right, inside from outside. Expressed directionally, a line can indicate motion forward and backward. A line takes on significance through its milieu, its relationship to context, the history of lines past, and the urgency of lines yet drawn. The question of what to call the drawing of such lines is a subject of intense investment: mapping, cartography, geographic information science, sketching, drafting, depicting, doodling, representing. Marianna Pavlovskaya views this contestation as productive. The work of GIS, as the drawing of lines, is largely undone: new versions, new applications, new contexts, new courses, new profit.

What is in the name "critical"? At times its use seems arbitrary. While at a workshop on the topic of qualitative GIS in Cardiff, Wales, in 2010, I discussed with a group of mixed-methods researchers the appropriateness of referring to their scholarly practice as "qualitative GIS" instead of the more widely known "critical GIS." The attachment of the word *critical* to their work served to generate unproductive or even negative attention. For many, *critical* was a word synonymous with *troublemaker*, and while they understood their scholarly practice to be a kind of methodological intervention, they recognized that the intervention of critical GIS was more obfuscating. That the name "qualitative GIS" would seem more direct and specific presents a quandary for the use of the word *critical* in mapping and GIS. Has this notion of criticality run out of steam?[1] Have there been too many mentions, too many invocations of its name, to make a difference today?[2]

At its worst criticality has become something of a joke—laughed about in graduate seminars and turned into a kind of highbrow drinking game at academic conferences. The joke is on all of us, though, as the stakes are high. Lost among the post-lecture receptions and cocktail parties is the sense of urgency that criticality implies (or used to). Not just a sense of accomplishment found in peeking behind the curtains to see the wizardry that produces our lives every day, criticality required patient observation, study, reflection, *and* a sense of responsibility to take action, to intervene, and to make change. Academic publication was only part of the means to a series of ends, not the end itself. Indeed, criticality has become ever more complicated and convoluted, as scholars assign great social and scholarly value to abstraction and detachment from or queering of public discourse. And while I would not suggest that criticality must be colloquial, I would argue that if we do not bring our various publics along in our thinking, then we have mistaken our responsibility for privilege.

However, I am getting ahead of myself. The point is that criticality—that is, being critical—is relational, and as such, is conditioned to change. I would not presume to set out a definition of *criticality*. Instead I offer a provocation for critical mapping scholarship: mimicking Latour, the tools of critique in mapping have not adjusted their targets as the techniques and technologies of mapping have evolved and proliferated. Put in the already dated lingo of software marketing: Where's the Critique 2.0 to go with Maps 2.0? How might we engage in postcritical mapping?[3] In this sense, there are limits to relying on 1980s–1990s critical perspectives on GIS and mapping for unpacking our contemporary moment. Times and technologies have changed. While age-old relationships between those that own the means of representation and those that are merely represented are still present, our renewed responsibility is to communicate the urgency of our perspective for publics thirsty for strong scholarship.

The stakes are high, and perhaps they always have been. Guy Debord understood this and challenged how one might conceive the map in relation to everyday life in Paris.

> The production of psychogeographical maps, or even the introduction of alterations such as more or less arbitrarily transposing maps of two different regions, can contribute to clarifying certain wanderings that

express not subordination to randomness but total insubordination to habitual influences.[4]

The point was to resist actions made habit through the relationship with the map. And this invitation to different conduct can even be felt in the revolutionary stirrings of the quantifiers of the same time period. In his preface to "Wild" Bill Bunge's revival of Fred Schaefer, Bill Warntz reminds us of the responsibility that comes with these postures:

> What's really involved is academic freedom. It has been said that a first rate college professor is always one who thinks otherwise. He [sic] is therefore doomed to be subjected to pressures from society, but that's understandable. Much more merciless and less comprehensible are those pressures that originate from within the academic community, especially from within his [sic] own discipline, initiated by those who wish somehow to keep the discipline "nice." Intellectual criticism developing from rival systems of thought is to be hailed. It is the truly suppressive action of those who do not "act" at all that hurts the most.[5]

While it may be considered sacrilege by some to read Debord alongside Warntz (the former a situationist and the latter a social physicist), the work made by these juxtapositions is to remind us of what we share in common and how we must recalibrate these collaborations.

Critical GIS—perhaps more than critical cartography—occupies the rub between such rival systems of thought. Steps removed from the visible signs of the representational production found in cartography, GIS advances a moment of further detachment from the author, the mapmaker. This presents a different crisis of representation, not only of the question of "the real" to which the map refers, but further to the question of by whom, by what means, the representing is even possible. In a commentary within *Political Geography,* cartographers and political military geographers rushed to the aid of the map, and in doing so, shamed geographers for gradually abandoning our most signature project. The culprit? That pesky criticality:

> A question that haunts critical cartography is does it really make good sense to analyze cartographic practices from the point of view of the crisis of representation in an age that witnesses an unprecedented expansion of surveillance, mapping and cartographic visualization in all walks of life.[6]

This is such a curious question to ask: "Does it really make good sense?" Why would the question of representation be nonsense in the rapid advance of location-aware technologies in everyday life? Criticality strikes a nerve, but the battles that ensue are ridiculously imprecise. Perhaps, instead of washing our hands of these transgressions, we need to take more responsibility.

Let us start again. How might we chart the changing valence of criticality, particularly in fields that intervene in the space created between theory and practice, such as critical GIS? Jeff Pruchnic summarizes this urgency for the broader field of critical inquiry:

> If the problem is not that critical theory's reliance on categories of oppositionality, resistance, and skepticism is in need of bolstering, but rather that these categories have turned out to be so powerful that they work even in the service of highly retrograde causes, those of us interested in the progressive possibilities of what we have come to call critical theory over the last several decades are left in something of a quandary.[7]

Therefore, what are the opportunities to revive and rethink the series of problematics that assemble critical energies around GIS in the 1990s and take new forms in Maps 2.0? What are the contours of practice that are absorbed within critical mapping, taking on new energy, and yet can work to actually unravel practice, through a kind of flattening?

In sketching the present history of criticality in mapping, I reflect on specific engagements that currently situate this criticality and outline the more pressing aspects of its research agenda. I trace a specific origin story for critical GIS (recognizing that there are alternatives), growing out of the GIS & Society tradition and responding to a longer-standing critical cartography. In doing so, our work is to continually reexamine those impulses and their forbearers, while recognizing the work that an origin story can do and has done for the field of critical mapping. An examination of the peculiarities of a critical impulse in mapping can begin to assemble a set of problematics that are uniquely responsive.

In the Beginning Was Friday Harbor

Origin stories both reveal and disguise. The force that motivates their telling is therefore always already a set of decisions about what was integral and what was marginal to the story, guided by hindsight. Or as Donna Haraway writes, "Epistemolophilia, the lusty search for knowledge of origins, is everywhere."[8] Stories of origins can strengthen bounds and bonds, secure territory, and identify inside from outside, friend from foe. For the merry band of scholars and map activists that currently invoke critical

GIS, a meeting at Friday Harbor, Washington, in late 1993 was the headwaters of the winding stream of criticality in the GISciences. In the beginning was Friday Harbor, so to speak.[9]

Around 1999 a specific solidification occurred. Nadine Schuurman's doctoral dissertation from the University of British Columbia was published in its entirety as an issue of *Cartographica*. This manner of publication was highly unusual, conspicuous, and fortunate,[10] marking a rising interest in the concerns of critical geographers as they turned toward GIS. She not only reviewed the recent history of conflict and contestation around GIS in the discipline, but—and perhaps most important—she articulated a position of situated critique, of care for the subject, that scribed value to learning the tools and the techniques.[11] She writes of her approach in the introductory chapter:

> This monograph is a vehicle towards creating a theory of GIS out of the ashes of criticism. It theorizes GIS as an emerging science by expanding the work initiated by the critics from human geography. . . . In the second half of the text . . . I move my focus to possible alternatives for critiquing GIS: ways of paying attention to both social and technical influences in the technology at the same time, thus interfering with the agenda of critics *and* proponents of GIS.[12]

This technopositionality (note her use of *both*, *and*, and *interference*) shifted how legitimacy could be produced.[13] The time had finally come for a renewed spirit of engagement.

The rise of GIS in geography departments in the 1980s brought about central questions about the role of GIS in reshaping geographical inquiry.[14] One such inciting debate was held between Peter Taylor, a political geographer, and Stan Openshaw, an acolyte of an emerging GIS. Taylor critiques the unchecked empiricism that GIS enables, where users "spend much of their time searching for problems for which they have the means to find solutions."[15] Openshaw fires back with a particularly vitriolic commentary, picking apart those geographers he characterizes as "parasitic" technophobes, practicing "second-rate philosophical thinking," unable to be or "do anything geographically relevant."[16] Taylor joins Michael Overton in direct response to Openshaw, and while much of the tussle is perhaps overly dramatic, at the core is a set of fundamental disagreements about how to socialize the practices of GIS—including the data central to such practices. Taylor and Overton plead for GIS to recognize its situatedness;[17] however, Openshaw remains uninterested in these kinds of questions and, instead of openly discussing the implications for his advocacy, decides to again cast his interrogators outside the fold, as "probably not geography."[18]

Of course this debate, if it could be called one, expanded to include other actors, bringing attention to the question of what kind of geography was to be furthered: those that directly engaged in militarism and violence, as discussed by Neil Smith,[19] or those that, in the words of Eric Sheppard, "narrow the variety of approaches to understanding the world."[20] As a result, a new direction in GIS scholarship emerged under the banner of "GIS & Society," propelled by a late 1993 meeting at Friday Harbor, a resultant edited publication called *Ground Truth*, and a special issue of *Cartography and Geographic Information Systems*.[21] These collections called for greater attention to the social implications of GIS and proposed the development of new practices that would be better attuned to these implications.

Of these new practices were two related movements: participatory GIS, which drew out the ways in which local, particularly indigenous, communities might utilize GIS to better articulate their concerns directed at more mainstream and state-based users,[22] and public participation GIS, which emerged alongside growth in public participation planning and sought to capture new participants in the use of GIS in planning contexts.[23] There are new challenges for these efforts, perhaps unforeseen in the late 1990s.[24]

Schuurman's monograph sought to advance a constructive science, pushing critics of GIS to respond in the terminologies of the technology itself and encouraging practitioners to better understand and attend to the implications of these tools. Critical GIS emerges as a field that actively pushes GIScience to incorporate a recognition of implications, while encouraging (although less so) social theorists to offer a more material critique.

This intervention created a discursive space in which a number of productive subfields emerge, including feminist and qualitative GIS. Central to these research areas was a concern for the incorporation of alternative forms of knowledge, as well as innovations in visualization techniques that might represent various intersectionalities. More recently, critical GIS scholars have turned toward the study of the proliferation of user-generated and other Internet-based content, part of what is termed the *geospatial web* (or geoweb). Drawing on interests in citizen science, a group of GIScientists (some of which are from the GIS & Society tradition) have pushed forward an agenda around volunteered geographic information, the scholarly twin of a more popular stream of activity in neogeography. Additionally, but not entirely unrelated, critical GIS scholars have been drawn into conversations with digital humanists, around what has been called the spatial humanities and geohumanities—taking up questions of the affordances of digital spatial methods for analyzing, representing, and interpreting an evolving humanities, amid the growth of digital information technologies.

Given these developments, the research agenda for critical GIS has never been broader. It has also remained largely unchanged since its earlier incarnation at the Initiative 19 meetings in 1996: the social history of GIS; the relevance of community-based GIS; issues of privacy, access, and ethics; the gendering of GIS; GIS and the environment; GIS and global change; and alternative technologies and alternative knowledges. Of course, central to this agenda is the recognition that GIS emerges from society, and that society, in turn, is influenced by GIS. Additionally, this agenda disentangles method from epistemology, recognizing that GIS is not necessarily reductive or positivist, while it lends itself to certain ways of knowing. Eric Sheppard, in his 2005 recap of this agenda, pushes critical GIS scholars to reexamine their relationship with critical theory and with the more conventional aspects of GIScience, the specific geographies of knowledge production and consumption, as well as the emerging science within their ranks.[25]

At the Revisiting Critical GIS workshop held at Friday Harbor during fall 2014, much of the original agenda was found to remain in play—although some concerns have evolved. In the wake of open data, volunteered geographic information, and neogeographic practices, there are also new and entrenched corporate players and renewed energies to surveil citizens, as well as more immediate methods to mark enemies of the state (for example, the rise of geospatial intelligence and human terrain mapping). I would summarize this persistent agenda in critical mapping as three advances: (1) disclosure and staging; (2) time and liveliness; and (3) doing, studying, and abstaining. This assumes a move from a more narrow set of concerns around desktop GIS software to digital mapping in the context of a pervasive digital culture.[26] I suggest that the implications of this move are titanic, as they place the GISciences within the realm of new media and communication technologies and a media studies form of cultural critique.[27] The signals of this move have long been present, if at the margins of the field. The time for taking this more seriously is upon us.

There are other origin stories to tell. These might include writers like Brian Harley and Nicholas Chrisman in the 1980s, or Howard Fisher in the 1960s and 1970s, or Guy Debord and Kevin Lynch in the 1950s and 1960s. Even the sensibilities of Richard Edes Harrison and Erwin Raisz or Edith Putnam Parker and J. K. Wright of the early twentieth century can be viewed as part of the story. (Admittedly these stories have a North American bias.[28]) Undoubtedly, the evocative conditions of mapping are fertile for criticality in various forms. Instead of creating division, why not work to widen the group and extend the rhythms of intervention and imagination?

Disclosure and Staging

More than making the map *do* something, criticality requires a different form of engagement. I lean on the map, which is to say that I burden it with the weight of important sociotechnical changes. The point is to thicken our understanding of the map, to think of it as more than a tool. This requires that we grant some agency, as James Corner might put it,[29] to the map as a convention that organizes and reflects both thought and action. More than a record of sociospatial phenomena, the map is also an artifact, a record of technocultural relationships.[30] Understood in this way, maps disclose an image of humanity in the moment of representation. Maps register the reverberations of these imaginations, of how humanity thinks of itself, from the Ptolemaic wind heads to the global inset maps in the margins. Historians of cartography would expect as much.

While perhaps no more geographically literate than our predecessors,[31] modern humans have never been more capable of using technologies to identify their individualized location on the surface of the earth. A dazzling mesh of wired and wireless infrastructure permeates the planet and reaches far into the atmosphere. These materials are recent manifestations within a long continuum of retentional techniques for the reproduction of humanity itself. This is my extension of an argument by philosopher Bernard Stiegler, a frequent commentator on the rapid pace of sociotechnical evolutions and revolutions.[32] By "retentional technique," I mean the technical practices in which humanity ensures its continuation.[33] These techniques include language and the spoken word, textual and iconographic artifacts, architecture and environment, and, indeed, the map.

The map stretches the lengths of this retentional continuum, albeit with differing rhythms and volumes of use. Beyond identifying *with* the land, humanity inscribes *on* the land with the broadest and most minute forces and movements—all made possible through the machines of our location-aware society. The map is therefore both a guide for and record of these processes of inscription. Indeed, as Corner argues:

> This is why mapping is never neutral, passive or without consequence; on the contrary, mapping is perhaps the most formative and creative act of any design process, first disclosing and then staging the conditions for the emergence of new realities.[34]

The process by which maps and other geographic representations disclose and stage requires more thought and experimentation.

More than a century of radical engagements with and on the map have made evident that the map is not only a window onto spatial phenomena. It is not *only* disclosure. The questions become: What might be the benefit of treating the map as a projection through which spaces are produced? How does the map *stage?* I tend to think of the map as an artifact of the times and spaces of map use, rather than a clarified vision of reality. Thought of in this way, the map is always already an externalization of human culture, memory, and action. The role of such a critical mapping perspective requires that we understand the reverberations of power that produced such maps and allowed them to persist. The question how does the map stage shifts slightly to who benefits from this staging? The beneficiaries of the map are but one example of the reverberations we might witness in our contemporary mapping practice. Indeed, there are other reverberations—technological advances, artist breakthroughs, and design achievements.

However, if Mercator maps cause radical cartographers to squirm with the uneasy legacies of colonial exploitation, how might our location-aware society today cause us to squirm, to reflect and react? I suggest that geodesign, quantified-self, neogeography, and big data are four examples of our new Mercators. These new mapping technologies are our digital pharmaka—both our poison and our most urgent cure, following Stiegler. These pharmaka reflect not only the reality we hope to understand and change but also the techniques for thinking about life itself. As such, these are retentional techniques—technical objects constituting both *to what* we pay attention, and there is much to which we must attend, and *how* we pay attention, and there are many demands on our attention. This power, to constitute attention, to stage, is what Stiegler calls psychopower. And while he limits his analysis to the telecommunications industry and its sublimation to advertising and marketing firms, I think we can extend this analysis toward new forms of spatial media, loosely grouped. These new Mercators have the ability to draw our eyes, to condition what we care about, how we might be convinced and form convictions, how we might act. Attention is care. And there is *much* to care about!

I suggest that by thinking of these four sociotechnical innovations as techniques to pay attention, where the map is a central figure, a new imperative to better understand the stakes and the importance of intervening is created. As I discussed in the introduction, the rise of big data science has created new platforms on which geography might assert and adjust public imagination. However, in addition to disciplinary restructuring, the impetus of big data has created the new coin of the realm, the new object of knowledge to be commodified: new forms of digital interaction, new forms of digital expression, new forms of digital profit. Undeniably, big data signals incredible ecologies—reterritorializing our thoughts and actions, as well as consuming earthly resources with renewed zeal.[35] There remains, nonetheless, a distinct possibility—however slim—to get it right, to use the mantra and capacities of big data to create more openings in discussion, more dialogue, and perhaps more just futures.

Within the GISciences, the rise of location-enabled big data has led to lines of inquiry around neogeography and the volunteering of geographic information.[36] As a subset of big data, neogeography describes more an affect, a sensibility, a playfulness with pervasive digital media. The imperative becomes to map yourselves and map your worlds. We are witnessing an enormous leap in the capacities of digital devices and platforms. Websites like Mapbox, CARTO, and ArcGIS Online offer unprecedented abilities to create visually stunning graphics and basic spatial analysis. We are told to create map stories and story maps, albeit on different websites, supporting different industries, different organizations. Indeed, there is intense competition for how to capture the laity, nonprofessionals and nonexperts. Eyeballs on mapping screens equal dollars. What messy alignments will be necessary to capture the energies and the attention of these emergent neogeographers?

While *neogeography* describes a specific activity of mapping enthusiasts, *quantified self* describes a creeping condition of everyday life. Many of us already carry on our persons some sort of hardware and software that captures streams of data about our movement and stasis. This moment of the pervasiveness of location-based technologies underscores the widening of the field. GIS indicates more than a particular piece of desktop software developed by Esri. Indeed, the great success of Esri is in the rapid advance of these tools into everyday life, from the desktop of experts into the pockets of users.

These previous three techniques—big data, neogeography, and quantified self—are reconfigurations of three conditions necessary for the rapid reorganization of everyday life. I suggest that big data operates as infrastructure, neogeography enables new labor relations, and quantified self enables methods of self- and collective governance. These technologies of our location-aware society are therefore not purely technical innovations but are evolutions in *technosocial* relations. It is within this perspective that I examine calls for geodesign, a relatively recent marketing term for a much longer effort in planning and decision making.[37]

Geodesign, short for geospatial design, is an attempt to solidify the software interests of landscape architects, urban planners, and architects. Beyond these more commercial motivations, geodesign calls us to better understand how the map *as design* serves to reconfigure time, "the future," through spatial experimentation. Esri touts the geodesign capabilities of their ArcGIS software, targeting not only the spatial relations of the present but also idealized and designed space-times of the future: "It gives us a new context for understanding, for moving beyond traditional mapping for navigation and location, and for using our maps for proactive designing and decision making." Their website continues, "Geodesign is our best hope for designing a better world."[38] This turn toward the future, to project future scenarios, occurs against the backdrop of social and environmental crisis—crafting a vision for more considered development amid looming apocalyptic ruin. At the center of these developments is the map, disclosing spatial conditions and relations in order to stage alternative realities, the one-two punch of our location-aware society.

Time and Liveliness

The map as an object and vehicle of engagement is always already more than it appears on the surface. There are movements—some subtle, others continental—that make the lines on the page and screen possible. To the map user, these lines appear as inevitabilities, nondecisions that emerge from the data untouched by opinion and taste. Indeed, this liveliness of the map just below the surface is at times difficult to apprehend and in other moments difficult to ignore. Criticality in mapping is partially about how to both apprehend (trace) and elicit (draw) this liveliness, to bring it back to the surface. Liveliness is therein an indicator for the distinctly critical capacity and the resonance of the map artifact. More conventional GIScience does not ignore this liveliness. Rather, the state of the science in mapping involves a series of confessions—the sins of the static map, the vanquishing of time by space, and the incantations of Hägerstrand and time geographies.[39] Herein lies the rub, the trouble of criticality in and through the map. Liveliness is an aspect of criticality in both mapping and more conventional GIS practices. The question becomes: what qualities of liveliness and what kinds of resonance?

In *How We Think*, N. Katherine Hayles suggests that the take-up of databases and computing by historians has renewed this multidecade critique of the representational limitations of GIS—namely, that space becomes fixed in the mapping of spatial phenomena. She contrasts the rigid spatial conceptualization offered by GIScientists with a concept of lively space offered by Doreen Massey in *For Space*,⁴⁰ depicting one well-storied wedge between critical geographers and spatial science. GIScientists sought methods to understand the spatiality of physical processes, but in so doing, as argued by critical geographers, they required that representations of space be largely inert. Meanwhile, critical geographers, in their quest to understand spatiality, felt they activated space by rendering it a *force* and not merely the *conditions* of relationality. Understood in this way, space was not just the chessboard but the series of relations, actions, and trajectories that compose the game of chess itself.

Somewhat sidestepping these more epistemological concerns, the subfield of spatial history,⁴¹ as argued by Hayles, revives these central questions directed at GIScience, namely, about how to handle time and, in doing so, avoid reducing space to a container. She writes of this distinction, as expressed in spatial history:

> Nevertheless, a crucial difference will likely always separate them as models for understanding the world: database technology relies for its power and ubiquity on the interoperability of databases, whereas narrative is tied to the specificities of individual speakers, complex agencies, and intentions only partially revealed. That is, narratives intrinsically resist the standardization that is the greatest strength of databases and a prime reason why they are arguably becoming the dominant cultural form.⁴²

Drawing on various projects from spatial historians, Hayles notes that although the relational database at the core of GIS would seem to enable alternative spatial conceptualization, time was reduced to snapshots, to static images of fixed movement.⁴³ Attaching points, lines, and polygons to narrative may actually delimit the kinds of narrative that can be advanced. In other words, placing history onto the map may only further distract from historical (and spatial) interpretation made most powerful in narrative.

Much of what is produced as spatial history could also be called qualitative GIS, but with an important distinction. Qualitative GIS attempted to advance qualitative forms of inquiry through the use of GIS. As a form of critique, early advocates of this approach sought to support alternative epistemologies within the terminology of the technology.⁴⁴ This meant adapting tools—even developing new tools—such that this inquiry would become possible.⁴⁵ Understandably, spatial history sought to leverage the spatial ontology and Cartesianism of GIS, requiring space to be fixed. The liveliness of relationships and events would merely dance across the surface. Space, as such, became background to time.

CRITICALITY

FIGURE 7. Enrique Chagoya, *Le Cannibale Moderniste* (1999). Mixed media on paper on linen, 48 1/8 × 96 1/8 × 2. U-5081-2002. Sheldon Museum of Art, University of Nebraska-Lincoln, Gift of Alexander Liberman and Frances Sheldon by exchange. Photo Copyright Sheldon Museum of Art. Reproduced with permission.

To understand the liveliness of space and the restrictive range of movement within traditional GIS, I tend to utilize art and art practice. When considering the potential of qualitative mapping techniques, I draw comparisons and creative energy from the work of Enrique Chagoya. His *Cannibale Moderniste,* shown in Figure 7, is my hope for qualitative GIS.[46] Time is not the primary domain of liveliness. Space is not merely a container, any more than the meaning of the painting is contained by the canvas. Instead the rub and rupture of space-times operate viscerally in this representation. The interaction of objects and concepts on the canvas is not seeking or producing equivalence or stability. Space-time is an opening. My point is that relying on a difference between databases and narrative masks a more fundamental, and resolutely geographical, question of spatiality. To engage in qualitative GIS—unlike spatial history—is to directly take up this question, not in the hope that an answer can be resolved, but that the pursuit of spatiality is our most urgent issue for the discipline of geography.

The map does not need to be silent on this question. (Nor should critical geographers feel they need to prove their mettle by eviscerating all mapping practices.) Indeed, mapping has always been an address to time, expressed as space. Whether scrawled on a napkin or produced through JavaScript, mapping addresses time, if perhaps to attempt to reduce the speed of time, to condition an opportunity to pause, reflect, perhaps change course. Criticality is to co-opt that address, such that space (including its expression) might be different. Of course, maps have a time; we often search out the date of a map, just as we might inquire about the date of a photograph. Further, maps make time, sutured to space. The hyphen in space-time highlights this urgency, to register a difference, to make action. As such, I read Guy Debord's maps of Paris and Bill Bunge's maps of Fitzgerald alongside more recent contributions like Lize Mogel's *Atlas of Radical Cartography* and the Critical Cartography Collective at the University of North Carolina–Chapel Hill. These are all mapping events that seek to challenge time through spatial expression, to activate space-times.

Instead of turning to spatial history to illustrate the ways in which the map bends toward the art of liveliness, my eye is continually drawn to the work of designers. The Urban Theory Lab (UTL) at the Harvard Graduate School of Design (GSD) is directed by Neil Brenner, geographer and political theorist. The remit of UTL is to interrogate systems of planetary urbanization.[47] Their book *Explosions/Implosions* provides a useful illustration for how mapping practices directly address space-time, to create conditions for liveliness. As part of the yearlong lab, student teams were assigned different parts of the planet, conventionally understood as nonurban. Research and drawing yielded a series of stories about the intensive and extensive operationalization of landscape for the project of planetary urbanization. The maps produced have a cinematic quality. Shots are framed, and the spatial viewpoint moves from the Cartesian to the oblique to the schema and plan. When they are most successful as visual strategies, they present undeniably the seemingly unchecked development of lands, seas, and atmospheres across earth—united toward a purpose for the creation and expansion of capitalism. These projects expand the range of techniques and tools for the expression of time-space. GIS, indeed even mapping, is just part of the process. The point is to resonate, to capture, *and* to catalyze an urgency.

Studying, Doing, Abstaining

"Yes, I *do* GIS." But this is to signal much more than the pushing of pixels across a screen. To *do* GIS is to engage a series of trajectories that crisscross and constitute digital mapping: the possibility of critique, a reactive technosocial history of the map, and the fashioning of spatial representations that sustain a progressive attention. The inseparability of studying the use of GIS and using GIS to study spatial phenomena is the particular technopositionality that I advocate in the drawing and tracing of lines.[48] In this sense, technopositionality might be best understood not so much as a method as an affliction. To tinker with the machine while investigating its provenance is such an example of a situated practice.[49]

Studying, doing, and knowing when to abstain is perhaps the more grounded concern of criticality for mapping in the location-aware society. For we can seek to understand how the map discloses and stages and how to engage a sense of liveliness in the ways the map addresses time, but *to map* one should conjure an ethics, a knowledge of what is appropriate, prudent, and most urgent. This is made all the more complicated as we broaden what is meant by *GIS*. The largely rhetorical move from GIS to digital mapping is a specific suggestion, in the hopes to revive the key tensions that gave rise to critical GIS while placing these tensions within a broader context of a pervasive and profit-driven digital culture.

To engage this technopositionality is to inhabit a field attempting to find itself. The University Consortium for GIScience (UCGIS) circled the wagons in 2006 and reinforced the mission of GIScience in a multiuniversity effort to describe the "body of knowledge" led by David DiBiase (now at Esri, formerly Penn State).[50] Of course, critical GIS is but a single unit of study, following ethics, of more than seventy units in ten "knowledge areas." This is an incredible maneuver, and the participants are to be congratulated, especially when considering the academic domains they attempted to bring to resolution (see Figure 8). The panel of experts delineated three domains of geographic information science and technology (GIST): (1) GIScience, (2) GIST applications, and (3) geospatial technology. The arrows have the effect of widening the field to the left and the right of the graphic, with a schizophrenic pole of geography versus information science, at the top and bottom of the graphic. Solid arrows indicate allies with reciprocity, such as that between geography and GIScience, while dashed arrows indicate asymmetries between fields—note that such a relation is indicated between GIScience and philosophy.

Viewed through the perspective of studying-doing-abstaining, the *Body of Knowledge (BoK)* is but a symptom of the broader trouble of the map in our location-aware society. There are so many connections, so many uses and users. Latour understood this quality of the map as an "immutable mobile." He writes, "All these charts, tables and trajectories are conveniently at hand and combinable at will, no matter whether they are twenty centuries old or a day old."[51] The map, as such an immutable mobile, is an artifact that provides a seemingly unassailable representation of reality (immutability) while also being endlessly transferable, mobile, and applicable in a variety of domains. The capacity of the map as an event that intersects so many domains presents further questions for the exercise represented by Figure 8. For instance, is geography meant to be the container

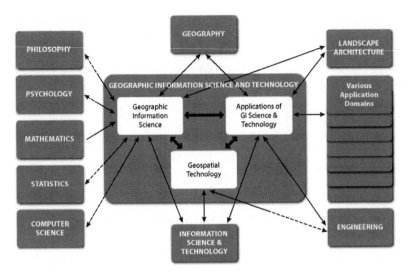

FIGURE 8. DiBiase et al., *Geographic Information Science and Technology*, 6. Image courtesy of David DiBiase. Copyright 2006 by the Association of American Geographers (AAG) and UCGIS. Reproduced with permission.

for the whole of social science inquiry into the implications of geospatial technologies? Furthermore, the arrows function as understated indictments of discipline. To engage in a project that takes up the technosociality of mapping, one must travel the solid and dashed lines between fields. For instance, to engage in philosophy, as in the creation of concepts, GIScientists must leave the workstation in the lab, knowing well that philosophers of the map will seldom travel that two-way path into GIScience. Herein lies yet more trouble.

To *do* GIS as studying-doing-abstaining is to always already question the assumptions of the mapping event, not to derail the endeavor but, quite the opposite, to ensure the resonance of the event. This might cause a rewiring of Figure 8, to bring the concerns of the unit of critical GIS out of the isolation of a single knowledge area in the *BoK*. To be resonant is to be responsible to the milieu of the mapping intervention. To be responsible is to seek to understand, to relentlessly study, that moment of representation.

To practice such a critical GIS, what ideally would be simply understood as *doing GIS*, is also to seek to understand in a most basic way when *not* to engage in mapping work. This aspect of technopositionality attempts to resist the map-or-be-mapped affect of much contemporary neogeographic impulses.[52] Abstaining seems radical, but it is a reaction to the ways in which even participatory mapping has enabled exploitation, as maps and local spatial knowledge become utilized in ways that exceed or even negate any simple emancipatory agenda. This should not be surprising as mapping is fundamentally capturing.[53] But this should not cause us *not* to map. Instead we should recognize the stakes and the need for being strategic in our mapping work.

This point, to strategically abstain from mapping, is underlined by the controversy of the Bowman Expeditions.[54] These "expeditions" (indeed, scare quotes are warranted) are named in honor of Isaiah Bowman, the geographer who advised wartime U.S. presidents Wilson and Roosevelt,[55] and are in part sponsored by the American Geographical Society and connected to the Foreign Military Studies Office. Professors Peter Herlihy and Jerome Dobson, from the University of Kansas, drew directly from the discourses of participatory mapping to engage indigenous communities of Oaxaca, Mexico, with a knowledge project consistent with American empire. In the wake of this unfortunate chapter of American neocolonial geography, we have an opportunity to be more precise in our discussions around what is meant by *participation, empowerment,* and *local knowledge.* To continue to borrow rather weak and problematic theories of participation from planning is to establish yet further grounds for exploitation and marginalization. Instead participation must be seen as just another configuration of power geometries. We should be able to recognize the radicality of being antiparticipation, antiempowerment. How might we understand such abstentions if we are narrowly focused on ascending metaphors of empowerment? Studying, doing, and abstaining from mapping is to understand our technopositionality within these geometries. As mapmakers and allies in mapmaking, we can never be outside these configurations.

As a third aspect of criticality in mapping, studying-doing-abstaining is a linked practice. It is a recognition, even an insistence, that ethics and criticality cannot be only a module attached to the last few days of a mapping curriculum, nor can it be a footnote to a sweeping hemispheric effort to enroll indigenous communities in a spatial knowledge project akin to dispossession. These new lines drawn and traced are not new in the sense of novelty but in the sense of an emergent, active connectivity, a creative process of walking new paths between domains, of a relentless search for reciprocity in practice and theory. Study, do, abstain.

Correspondence

If we conceptualize criticality as a force in tension with technicality in mapping, critical approaches have some steep competition with more technical pursuits. Stroke widths, object alignments, and typographic stylization dominate the fine last-minute adjustments made in illustration software during mapmaking. The immediacy of computer-aided mapping provides nearly instant gratification, while the relatively slow, patient work to build capacity for critical inquiry realizes itself at more extended rhythms. In other words, consider the difference between drawing a line and reading (the histories) of a drawn line. In this respect, I find it useful to consider the more intensive and extensive labors of mapmaking nearly a century prior. While contemporary mapping works likely share few practices in common with their predecessors, that moment of experimental tinkering with a drawing binds these practices from different times together. To conclude this chapter, I briefly ruminate on one of the first computer maps, which I articulate as part of a rich continuum of early twentieth-century experimental methods in mapmaking. What is the role of such experimentation in geographic representation today? How might we register the reverberations of such work as part of a critical endeavor?

The ability to see the world as it unfolds around us is inherent in the mapmaker's craft. Two figures in the history of cartography, Erwin Raisz (1893–1968) and Howard Fisher (1903–1979), among many others, employed different techniques to shape what was possible in mapmaking, to ensure correspondence between the representation and reality. Consider Raisz's work with map projections in 1943:

> Although the distortions on the peripheries of the orthographic view may be extreme, we perceive the correct proportions because we visualize a three-dimensional body instead of a flat map. The use of these projections is limited to illustrative and educational maps, and in many cases the results will be amusing rather than of practical value, yet they open up a new avenue of experimentation, the end of which has not yet been seen.[56]

For Raisz, the work of geographic representation was considered largely perceptual—to view the entire planet at once, while communicating a correspondence with the round Earth. His "armadillo" world map is one of a series of experimentations with orthoapsidal projections (he also experimented with lima bean and scallop muses).

While he would witness the beginning of a new mapmaking method at Harvard led by Fisher, Raisz's practice would solidify geographic representation in traditional, manual methods.[57] Nearly forty years after Raisz's hand-born work on map projections, Fisher would similarly contemplate this issue of comprehension and correspondence:

> It will usually be necessary for the map designer to make a conscious choice between generalization and particularization. Each has its value, but if the main goal is good overall comprehension, we will usually need to generalize to a substantial degree.... No method has ever been invented which is capable of optimally serving both goals simultaneously.[58]

These quotes connect two experimental processes of crafting a vision of the world. More than a method, then, Fisher's Lab for Computer Graphics (LCG) at Harvard continued a particular make-do attitude in the 1960s advent of the digital map—to take what was perfectly adequate in one domain and apply it, make more of it, in another domain. Fisher and his team of programmers decided to use an overprinting of characters to produce visual densities. Central to this kind of digital experimentation was an interest in the projection of a three-dimensional image across a flat medium—an interest well established in the history of cartographic expression.

This experimental spirit at Harvard was born from a series of events that connect the campus of the mid-1960s to the broader revolutionary stirrings in the spatial sciences across North America.[59] One of these mid-century revolutionaries, Bill Bunge, discussed this emerging approach to analyzing complex social and spatial phenomena, and the representation of such analyses, in his dissertation:

> The exact distinction between the map as a logical system and the map as a framework for theory is murky, but it is of the utmost importance since theory can only be as powerful as the underlying logic. For this reason cartography is placed in company with mathematics.[60]

For these geographers, the map was burdened by both the particularity of observable facts and the theorization of geographic phenomena (here understood as map abstraction or generalization). An epistemological turn was afoot, driven by the use of the computer in spatial analysis. A new language to express geographic relationships required a revolutionary method for representing these relationships.

Like Raisz's tinkering with map projections in prior decades, these experimentations intended to expand the possibility of observing a changing planet. Far from requiring scientific expertise in interpreting geographic representations, these drawings sought to make the world viewable—and therefore accessible—to a public eager to understand the planet's spatial dynamics. The computer, for Fisher, did change the game in thematic cartography. Perhaps more important, however, the new capabilities offered by the digital map raised concerns for general comprehensibility.

In the fifty years since the first digital maps, the practices of geographic representation have changed both in speed and in volume, whereas the model of map communication and interaction has remained largely the same. Maps are envisioned as documents that, when designed with great efficiency, effectively communicate information about a variety of spatial phenomena. They correspond. As a method to collect, analyze, and represent geographies of information, mapmaking is burdened by the weight of these models of correspondence and comprehensibility, establishing a particular register within which information about the world is to be viewed. Contemporary techniques in geographic representation, sometimes termed geovisualizations, consider the map as perhaps just another informational graphic.

Again, it is appropriate to reconsider those same interests in mapmaking of nearly a century ago. Richard Edes Harrison (1901–1994), a contemporary of Raisz, sought to break with cartographic dogma, to introduce different perspectives on the planet, to create correspondence that would be undeniable by the public. Susan Schulten reflects on her interview with "Rikky":

> When I interviewed Harrison in New York at the end of his life in 1993, he still insisted that I call him an artist rather than a cartographer, for he disdained the constricted techniques of mapmakers who were hidebound by convention.[61]

His approach was featured in *Life* in 1944, with a six-step process of producing wartime visualizations of continents.[62] By staging a photograph of a globe, Harrison created a fictional, aerial viewpoint, "40,000 miles over the middle of [the] Atlantic Ocean," from which readers could imagine the geographical complexity of world war.[63]

Certainly, then and now, the drawing of a line is the making of a difference. However, the gloss, or patina, of contemporary geovisualizations, made with advanced geospatial and illustration software, can risk breaking with a key commitment of early to mid-twentieth-century geographic representations: a commitment to the wide participation of a map-reading audience in the reading and wonder of the world. In works by Raisz, Fisher, and Harrison, the map, as a vehicle for representation, was a kind of fantasy, a creation. The map was vernacular yet artisanal, grounded yet imaginative. The point was to see the curved planet, to understand the undulation of the landscape, to imagine themes of daily life that cannot be physically seen but are experienced nonetheless.

These experiments in mapmaking mark a culturally conditioned comprehensibility, where correspondence between reality and map pulls the map reader more deeply into the representation, to make interpretations and draw strong conclusions. These experimentations, to "make do" with available techniques, should inspire contemporary mapmaking. How might design resist the gloss of spectacle and elevate slow-mapping, where the representation intervenes in the known? How might the fine adjustments made in geospatial and illustration software further disguise the mechanics of representation from a public? How might cartographic experimentation forgo the rush toward a faddish polishing of infographics and instead amplify the disruptive potential of geographic representation?

CHAPTER TWO

Digitality

Origins, or the Stories We Tell Ourselves

It's very late, but just perhaps not too late to be decent.
—WILLIAM WARNTZ, "PREFACE"

This disassociation between an art and its history is always ruinous.
—GILLES DELEUZE, *TWO REGIMES OF MADNESS*

The *Harvard Papers in Theoretical Geography* were edited by Bill Warntz (1922–88), the second director of the Harvard Laboratory for Computer Graphics and Spatial Analysis (LCGSA). He opens the papers in 1968 with an introduction to Bill Bunge's (1928–2013) manuscript on spatial analysis and the trials of Fred Schaefer (1904–53), a tribute to the deceased geographer from the University of Iowa.[1] Schaefer advocated a scientific approach to geography, and his *Exceptionalism in Geography*, published posthumously in 1953, was a momentous attempt to critique an unchallenged prioritization of the region in spatial inquiry.[2] Bunge's manuscript on Schaefer has the feel of an obituary with a biting, impassioned edge. Warntz's introduction describes the difficulty with which Bunge was able to publish his research, paralleling the difficulties Schaefer experienced in publishing his *Exceptionalism*, a couple of decades earlier. To begin the *Harvard Papers* with controversy underlines the affective moment of 1960s American geography. Areas of specialty were being reconfigured. Departments were being reorganized with the emergence of new computing power. So perhaps it is also very late to tell the story of another key figure from this time period, often lost in the shuffle: Howard Fisher. His efforts, some intentional and others happenstance, condition the beginnings of the digital map as we interact with it today.[3] If the tracing of maps is a key part of learning the craft of hand-drawn cartography, how might we trace the digital map?

To document and attempt to "map the trace" of the digital map is to also engage the proliferation of contemporary applications of the digital map. As both a material and discursive artifact, the digital map is more than a map presented through digital technologies. Rather, the digital map marks a distinct form of interaction with representation. The digital map, as an exemplary artifact of technoscientific achievements in the GISciences, has had diverse impacts throughout society, from the more obvious (smart warfare, just-in-time manufacturing information systems, and so on) to the more everyday (wayfinding and navigation, tracking of coded objects, and so on). Understood as the social implications of geospatial technologies, this form of scholarship (and tech journalism) is well trod.[4]

To tell these other stories, we could also examine the ways in which industry represents itself. Advertisements and other marketing strategies dominate the inboxes of mapping enthusiasts with new corporate players, such as Mapbox, CARTO, and Mango, which have entered the crowded space dominated by Esri and Google. These strategies document the emergence of a particular kind of mapping practice (crowdsourced, open-sourced, interactive, mobile first, emergency ready) that is resolved through specific technologies. For example, consider the website for Esri's Disaster Response Program shown in Figure 9. The intent of Esri's strategy is clear—to capture the market for geospatial technology amid increasing social and environment instability: "Help when you need it most . . . as part of our

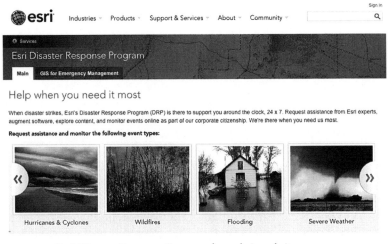

FIGURE 9. Esri Disaster Response Program from their website.

corporate citizenship."⁵ To pay attention to these marketing attempts is to map the trace of industry and innovation in mapping.⁶ Donna Haraway is my guide:

> I am reminded of David Harvey's (1989: 63) observation that advertising is the official art of capitalism. Advertising also captures the paradigmatic qualities of democracy in the narratives of life itself. Finally, advertising and the creation of value are close twins in the New World Order, Inc.⁷

Corporate citizenship in the New World Order, Inc. indeed. Our map of the trace of the digital map always already begins from our contemporary moment, where the organization of everyday life is structured by geospatial industries, where that structuring is meant to constitute an affectual relation through the tools of marketing and advertising.

However, the stories associated with the beginning of the digital map leave other traces. In 1938 Erwin Raisz stated what would become obvious, if opaque, "Maps constitute an important part of the equipment of modern civilization."⁸ By the early 1960s, Raisz would be in attendance at luncheons at Harvard's Faculty Club organized by Howard Fisher on the topic of computer mapmaking. Indeed, our well-told origin stories associated with GIS focus there and then, around experiments in computer-supported cartography at the Laboratory for Computer Graphics (and Spatial Analysis) founded by Fisher in 1965. Stories of the history of the digital map emphasize the lab and, in particular, the productivity of *men* at the LCGSA. (A full history of the social reproduction that produced the digital map has yet to be written, although work by Judith Tyner and Will van den Hoonaard are notable and needed interruptions.⁹)

To map the trace of the digital map, a project I suggest is open and unending, I place my archival work with the Fisher papers among many other trajectories that situate this time-space of American geography. While tracing the correspondence of midcentury scholars in computer-supported cartography, I read the historical material manifestation of the closed world and more specifically the rise of social physics and histories of the so-called quantitative revolution.¹⁰ But there are other stories, of the closure of academic departments, the project of American empire, the alternative and radical practices both within the discipline and without.¹¹ In what follows, I lean on the Fisher papers to understand the provenance of the LCGSA, the specific problems that were meant to be solved, and the productivity of experimentation in cartography, while proposing lines of inquiry that enlarge what digitality would come to mean. The point is to thicken the digital lines drawn with computation, recognizing these moments as technosocial—not in order to constitute a rigid intellectual history but to disrupt easy origin stories with the cul-de-sacs of experimentation and failure, tenuous allies and adversaries, and the fragility of thought and action.

Problem Solving

Before introducing the biographical detail of Howard Fisher, I like to imagine him in his freshmen seminar classroom in Harvard Yard in 1966. The seminar was focused on problem solving, and this theme is no better observed than through an activity he designed for these freshmen, encouraged by similar activities Fisher had developed while at Northwestern University. The activity begins:

> Here are five problems rolled into one. Of varying degrees of difficulty, they are representative of a wide variety of problem situations. While we will, of course, be interested in your specific solutions to each, we will be especially eager to have your advice as to superior methods of attack. Accordingly, note carefully your own mental processes as you solve the problems.[12]

One of these problems asked students to brainstorm as many answers as possible within twelve minutes, spurring them on with competition from the results of a similar activity at Yale. The students were shown a drawn image of a human hand—with an extra thumb:

> We don't think this is very likely to happen, but imagine for a moment what would happen if everyone born after 1966 had an extra thumb on each hand. . . . Now the question is: What practical benefits or difficulties will arise when people start having this extra thumb?
> . . . Working as fast as possible, write down all possible ideas as they come to you—including wild ones.[13]

At the conclusion of these ideation sessions, Fisher would then ask his pupils to reflect and record how they thought about the problem. Of course, Fisher's interest was not in the delineation of a variety of ideas for the use of an additional, opposable digit, but in the ways in which complex and creative solutions were thought.

I suggest that this orientation toward the task of mapping made him both an innovator and an agitator within the field of cartography. The arrival of new computing systems on university campuses meant new capacity. Fisher was disinterested in the routine utility of these systems, but interested in how they might be pushed to do something not yet imagined, how to use and abuse these techniques in service of something more disruptive. Cartography, a relatively recent science, had passed through the military-industrial machinery of the postwar period. Traditions were revaluated in place of standardizations of (machine-supported) hand-drawn cartography.[14]

Fisher was born in 1903 and died in early 1979. After earning his bachelor of science degree at Harvard in 1926, and spending just a year studying architecture, he moved to Chicago to found a company called General Houses, Inc. He set about creating solutions to the problem of housing, adapting factory methods to the development of prefabricated housing, concrete curtain wall systems, and factory-made and site-assembled stair systems. Unfortunately, the Great Depression would mean that General Houses would build and sell few prefabricated houses. Fisher instead found a variety of work consulting on shopping centers and freeway bypasses as a new automobility swept municipal planning and development.

His collision with the use of the computer for mapmaking would happen at Northwestern University, where he had been a lecturer since 1957. And so the story goes that in 1964, Fisher enlisted the support of Betty Benson (1924–2008), a computer programmer, to develop the Synagraphic Mapping System (SYMAP).[15] This work captured the attention of those at the Harvard GSD and Dean Josep Lluís Sert. At Harvard, Fisher would question some of the assumptions of midcentury cartography. He attempted to rigorously understand the impact of the computer on thematic cartography, and with these new innovations, he experimented with the various conventions of subject selection and generalization.

His approach ruffled feathers, as can be seen in this transcribed conference session from 1970 between Fisher, George Jenks, of the University of Kansas, and Arthur Robinson, of the University of Wisconsin:

> PROFESSOR FISHER: You are the leading expert on this, and we don't know as much as you. As a result of your comments, we will go back and run more [computations] and see how we like it.
> PROFESSOR JENKS: Good.
> CHAIRMAN ROBINSON: I think we have taken care of that subject.[16]

While only an excerpt, we can sense the tension between the cartographic establishment represented by Jenks and Robinson and the new computer cartography. Fisher defers to Jenks on the topic of generalization. Jenks's curt reply is followed by Robinson, who hopes to move the conversation along. In these moments, Fisher's first reaction is not to stubbornly dig in but to lean on a relentless empiricism and experimentation. Contra Robinson, the subject of map generalization and the persistence of data were far from handled.

University Subjects

I have already moved too quickly in this story. Return to 1930, the year Fisher founded General Houses, Inc. in Chicago. In that year at Harvard, Alexander Hamilton Rice (1875–1956) entered into an agreement with President Lowell. This document indicated that Rice was "desirous of giving concrete expression to his interest in the development and future of this science through the establishment as part of Harvard University of a School of Geography." The agreement continues to note that Harvard "recognizes the far-reaching benefits to the science of geography." Far-reaching indeed. The agreement established the intended focus of the school: "the teaching of fundamentals of geographical science and their application to pure regional geography, particularizing, especially, in geographic and physiographic problems and their relation with complementary ontographic problems." The school would be built and operated through funds from Rice, and his gifts would continue as long as Rice was able and interested, and that, "upon his death or retirement" (of importance to this story), such funds would ensure the continuation of geographical work within the school. The two-page agreement was signed by both Rice and Lowell on June 18, 1930.[17]

It is generally well known that much of Rice's fortune was from Eleanor Widener, who is weaved into stories told on Harvard Yard connecting the monstrosity of the *Titanic* with that of Widener Memorial Library, which opened in 1915, to memorialize Widener's son and husband, who perished with the sinking of the British passenger liner. Indeed, it was reportedly at the dedication ceremony of the building on June 24 that Rice met Widener. A relationship blossomed. Widener's fortune would support Rice's interest in the building of the Institute of Geographical Exploration on Divinity Avenue, just outside the Yard walls, and geographers took roost. Pure regional geography would have a home, and the chief cartographer would be Erwin Raisz, who was introduced to Rice by William Morris Davis.[18]

Raisz is credited with the first general cartography textbook of the modern era in English, published in 1938.[19] His students were trained in pen and ink cartography, and his drawings—including his maps—highlight a hand-born craft in its final moments of the twentieth century. Figure 10 shows a section of his most famous map, *Landforms of the United States*.[20] In this photograph of the original manuscript, you can see edits that Raisz had made in this sixth edition from 1952. Raisz was committed to a democratization of geographic representation, with maps and other graphics that could be read at the surface without a belabored expertise.[21] This drawing technique produces a style that is unmistakably Raisz, especially

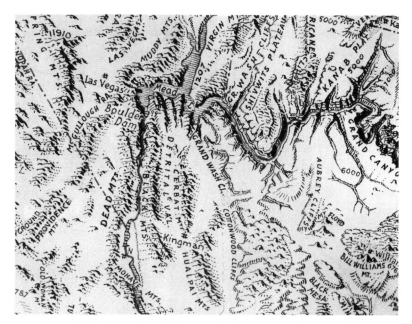

FIGURE 10. Enlarged section of Raisz's original manuscript for *Map of the Landforms of the United States*, 1939 (6th rev. ed., 1952). Image courtesy of Lize Mogel, The Harvard Map Collection, and Raisz Maps.

when compared with contour-line or relief-shading methods for representing terrain.[22]

Raisz and Hamilton Rice's twilight years would see the demise of the institute and the scattering of geographers from Harvard, in a story that borders on lore. Neil Smith relates the tale of two university presidents, Isaiah Bowman, geographer and then president of Johns Hopkins, and president of Harvard Jim Conant.[23] But before Conant would speak those words in 1948, that geography was not a university discipline, there would be discussions, Smith argues, of homophobia and an "unsavory" relationship between the chair of the department, Derwent Whittlesey, and his friend and lecturer in the department, Harold Kemp.

Geography at Harvard came to an abrupt end in 1948.[24] Furious, Rice pulled his endowment arrangements, just eight years before his death in 1956 would have perhaps made the gift permanent. In the same year, Whittlesey's death would leave Harvard without a geography professor and coursework.[25] Indeed, from 1956 until 1964, one might argue that little if any geographical work was explicitly underway at Harvard.

Harvard's loss would be quickly overshadowed by revolutionary stirrings at the University of Washington, where a new cohort of geography graduate students arrived to study with Edward Ullman, a former geography professor at Harvard who had left in 1951. Geographers like Dick Morrill, Duane Marble, and John Nystuen were drawn to Ullman.[26] However, as Ullman was away on fellowship, many took up study with Bill Garrison—who introduced them to statistical methods in mapping. This rubbed against an established regional geography in the discipline, igniting what is now fabled as the quantitative revolution. Geographers from the University of Washington, including Waldo Tobler and Bill Bunge, added fuel to the fire.[27]

For these "Garrison Raiders," geography was "a basic science." Bunge continues in the concluding statement of his PhD thesis, "Theoretical Geography," "It has an excellence of its own. Its spatial point of view leads to questions that produce knowledge peculiar to it among all the sciences."[28] Cartography was considered the vehicle for this revolution. Bunge writes:

> The reason geography has always paid such respect to maps is that they have been the logical framework upon which geographers have constructed geographic theory. They have been to geography what mathematics has been to some other disciplines.[29]

Caught between a discipline somewhat spellbound by generations of regional geographic inquiry and the new statistical and computational techniques of the day, Bunge's thesis grapples with what might be meant by theory—and the role of mapmaking in relation to theorization.

Bunge found that geography's claim to be a science would require more explicit treatment of the relationship between logic (the domain of mathematics) and theory. He explains further:

> Which, then, is cartography, theory or logic? . . . If we say a^2 plus b^2 equals c^2 without identifying the a, b, and c with observable phenomena, then we are dealing with pure mathematics, a system of logic, a deduced system of relationships. But if we identify the a, b, and c with some observable phenomena, say, numbers of apples, oranges and peaches, then the formula is a theoretical statement. In similar fashion it might be argued that the map is capable of portraying the spatial property *shape* without reference to any observable shape. But if we map the outline of Long Island, we are dealing with theory, since we have identified the abstraction *shape* with a particular set of observable facts, the outline of Long Island.[30]

At the dawn of the 1960s, Bunge's work provides a useful mile marker, to understand how a series of transformations would occur in the conceptualization of the map. Certainly, the role of the map was integral—not only as representation of geographic phenomena or illustration of regional differentiation but as a kind of logic. In so doing, Bunge suggests that the map would need to become the vehicle for geography's rise to the level of science. He underlines the significance of this move:

> The methodology endorsed here is one which leaves geography no excuse for not reaching full maturity as a science. While nothing in present-day geography is discarded, the overall importance of theory and the inevitability of mathematization is insisted upon. By identifying regional geography with facts (description), systematic geography with theoretical geography, and cartography with mathematics, the following arrangement of the discipline appears: [See Figure 11].[31]

Nothing would be discarded. Regional and systematic geography would rest comfortably with cartography. That this well-reasoned sentiment was advanced by a doctoral student at the beginning of his scholarly career is made poignant by the difficulty Bunge would face in getting this line of thinking published.

Regardless, Bunge and these Washington graduate students lived this particular relationship. Their experiments with computer-based cartography allowed them to establish a kind of logic to geographic theory and observable facts. Alongside these experiments and conceptualizations came innovations in computer mapping. For instance, in a paper from 1959, Waldo Tobler explains a method of using 343 punch cards to plot a map of the contiguous United States in a lightning-fast fifteen minutes.[32] And they were learning these methods because many of the grad students had funding to work in the Washington civil engineering department, under Edgar M. Horwood (1919–85), who was awash in grant money.

Horwood, who trained at Penn under Lewis Mumford, was exploring computational methods to study urban blight. Figure 12 demonstrates the Horwood method, created with the energy and innovations of the Garrison Raiders. Such a field of numbers represented blight by census block in Spokane, Washington, and is, in my opinion, one of the first leaps in the

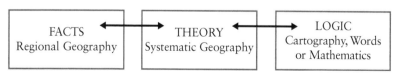

FIGURE 11. Bunge's configuration of the discipline of geography in "Theoretical Geography."

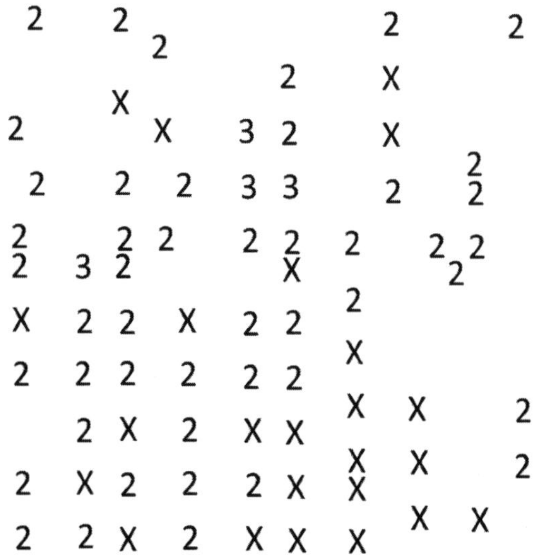

```
    2   2                   2       2
            2
                    2   X
    2       X   X   3 2   X
        2   2   2   3 3     2   2
                                2
    2   2 2         2 2
    2   3 2   2     2 X   2   2 2
                    X         2 2
    X   2 2   X     2 2   2
                          X
    2   2 2   2     2 2
                          X X         2
        2 X   2     X X   X X
                          X
    2   X 2   2     2 X   X           2
    2   2 X   2     X X   X   X X
```

FIGURE 12. Enlarged and redrawn section of Ed Horwood's 1962 digital map of urban blight in Spokane by census block.

rise of the digital map.[33] Horwood traveled the country in those early years of the 1960s, sharing this computer-mapping method. One such workshop in the summer of 1963 was held at Northwestern University, on the topic of computer applications for urban analysis. At this workshop, Horwood met Howard Fisher. Fisher, impressed by the possibility of this method, but not of its readability, took to innovating on the approach.

SYMAP

In early 1964 Betty Benson worked with Howard Fisher at Northwestern to program SYMAP, even giving it its nickname.[34] SYMAP employed a method of overprinting symbolism that would make Horwood's printouts easier to comprehend. A press release, "Computer Mapping Technique Developed at Northwestern," was released December 9, 1964, no doubt amid concerns that the origin story of the system would get muddled by Fisher's move to Harvard:

> Fisher said the process may prove useful to city officials, planners, urban and real estate analysts, or businessmen needing maps—especially maps which are expensive or complicated when drawn by hand. . . . "A map which might cost $200 to draw by hand could cost as little as $10 to do by computer," Fisher said. Its production time

> could be reduced from a week or more to a few hours, he said, with actual computer time amounting to no more than a minute or two. . . . Using SYMAP, Fisher said it is possible to produce complex maps which are not only geographically more accurate, but also illustrate such things as residential housing conditions in an area, assessed valuations of real estate, crime statistics, health statistics, market information, and a wide variety of social and economic data.[35]

Figure 13 demonstrates the use of this overprinting process, what former users of the program might recall as OXAV.[36] While there were more advanced computing machines and output options, Fisher's vision was committed to using machinery that made digital map "mass production at different computing centers" possible nearly "anywhere."[37] Densities and surfaces were produced in ways that could be read by a novice map reader.

After visiting in 1964, Fisher arrived permanently at Harvard and formed the Laboratory for Computer Graphics, and made a proposal to the Ford Foundation in September 1965 for which Harvard was awarded nearly $300,000. The goal of such a lab was

> to help raise the performance level of professional persons in city and regional planning and related fields through the more extensive and more sophisticated use of factual information—as now increasingly made possible by the expanding field of computer science in combination with advanced statistical techniques, systems analysis, and similar analytical and decision-making procedures.[38]

Number of intervals specified	Standard symbolism for intervals 1	2	3	4	5	6	7	8	9	10
10	.	'	-	=	+	X	O	Θ	⊠	■
9	.	'	=	+	X	O	Θ	⊠	■	
8	.	'	+	X	O	Θ	⊠	■		
7	.	'	+	X	Θ	⊠	■			
6	.	+	X	Θ	⊠	■				
5	.	+	O	Θ	■					
4	.	+	O	■						
3	.	O	■							
2	.	■								
1	.					(OXAV=■)				

FIGURE 13. OXAV symbolism for SYMAP data class intervals, in the 1975 SYMAP User's Reference Manual by James A. Dougenik and David E. Sheehan. Harvard University Laboratory for Computer Graphics and Spatial Analysis. Image courtesy of Harvard University Archives.

The stories that surround the lab typically center on such figures as Jack Dangermond, who was recruited by Fisher and who earned his MLA at Harvard in 1969 before forming Environmental Systems Research Institute (Esri) with his wife, Laura, in Redlands, California. In the early days of his consultancy, Dangermond used and developed the SYMAP program further, and he would go on to recruit members of the lab to join operations at Esri.

Of course, the story of the lab in the 1960s is more crowded and complicated, with individuals like Robert Weinberg, a GSD alumnus, who gave funds that launched the lab in 1965 and ensured its continuation in 1974 when confidence levels were low at the GSD. Don Shepard, now professor at Brandeis, was a freshman in Fisher's first Harvard seminar in problem solving in 1966 and responsible for a significant mathematical innovation in the problem of spatial interpolation. One stipulation of the Ford Foundation grant was broader dissemination of the approach, including a correspondence program that in 1967 had more than five hundred participants around the world and consumed significant lab resources, largely staffed by Helen Mansfield and Marie Gardner. The lab was quite well funded, despite a perpetually tense relationship with the GSD.

William Warntz joined in 1966. Previously at the American Geographical Society in New York City with training as an economic geographer at Penn, Warntz had worked with John Q. Stewart at Princeton to advance a spatial social physics. Warntz was named professor of theoretical geography, although he did not hold a chair at Harvard—a significant distinction. He was the first professor of geography at Harvard since the death of Whittlesey, ten years prior. By the time he arrived in Cambridge, Warntz was well into his scholarship on surfaces, and the *Harvard Papers in Theoretical Geography* bear this imprint. Allan Schmidt (1935–) was recruited to the lab by Fisher in 1967. Schmidt was at Michigan State working on a program called METROPOLIS, a kind of urban simulation routine. He had used SYMAP to create an animated cartography of Lansing, Michigan.[39] Fisher put him in place as associate director of the lab, which, with Warntz, became the Laboratory for Computer Graphics and Spatial Analysis.

Fisher pushed Horwood to recommend someone to spend a few months in Cambridge, without results. He recognized the pitfalls of engaging in methodological development without disciplinary mooring—that the lack of a formal geographic foundation colored the contributions of the lab within those communities most directly hailed by the lab's work.[40] Despite a spending spree to attract as many geographers as he could (the Ford Foundation required that the entirety of the grant should be spent by 1969), Fisher understood that the field of quantitative geography was rapidly changing and he would need to get up to speed on the state of the art. Initially this meant transferring the books—and as a result the local stores of geographic knowledge at Harvard—that used to be in the geography library over to the lab in its location in the basement of Memorial Hall, room 121. This also meant getting caught up on the work of Bill Garrison and his geography graduate students at the University of Washington, although doing so without raising too many eyebrows. Fisher gave instructions to an administrator at the lab:

> Buy book *Quantitative Geography,* William L. Garrison, Ed. New York. Send letter to secretary [of] Dr. Garrison in Department of Geography, Northwestern University. Ask her to tell us how we can order one of these books so as to get it as soon as it appears. Leave Fisher's name out entirely. Then order a copy as soon as it is issued.[41]

It is easy to forget the importance of written scholarship to the advancement of ideas in the academy. That Fisher's name was to be left out of the correspondence should perhaps be understood in the context of competition in a rapidly evolving field. Fisher also reached out to one of Garrison's graduates, Waldo Tobler, who had established an academic career at the University of Michigan and was working to innovate his computer-mapping method. Tobler recommended a number of key texts of the day to Fisher, including Peter Haggett's *Locational Analysis in Human Geography* (1965), Arthur Robinson's second edition of *Elements of Cartography* (1960), and Ed Imhof's *Kartographische Geländedarstellung* (1965).[42] Fisher immediately set his staff to acquiring them.

Fisher's practice of correspondence was unrelenting, placing him in touch with many of the contemporary thinkers in academic geography, including Robinson at the University of Wisconsin. In a letter to Robinson in 1966, Fisher writes, "We are primarily concerned with the goal of increasing the communication value of the type of maps we are producing."[43] The "communication value" of SYMAP (shown in Figure 14[44]) was likely in question by Robinson and other cartographers of the day, such as George Jenks at the University of Kansas. However, Fisher felt that in time these prickly issues would fade—that with greater experimentation with and exposure to computer mapping methods, cartographers would broaden and deepen their impact. For him, decisions about the symbols used to represent a statistical surface were often left to aesthetics, instead of a

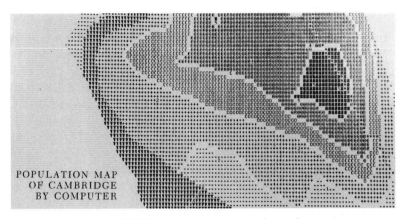

FIGURE 14. Section of SYMAP output showing population density in Cambridge, created in 1966 by James A. Dougenik and David E. Sheehan. Harvard University Laboratory for Computer Graphics and Spatial Analysis. Image courtesy of Harvard University Archives.

more data-led method. At the 1970 conference at Northwestern, David Sheehan (a member of the lab) gets into an argument with Robinson:

> MR. SHEEHAN: . . . As I said, I think geographers tend to look towards map aesthetics more than they do towards data. . . .
>
> PROFESSOR ROBINSON: I think what I want to object to is the term "aesthetics." There is a tremendous amount of information in a dot map because the dot map by itself is countable. The densities are not easily calculated in one's head, but that doesn't take away the fact that there is a tremendous locational value in a dot map. . . .
>
> MR. SHEEHAN: . . . As I said, otherwise, I think it is time that dot maps graduated, so to speak. They are at, basically, a kindergarten stage, . . . I think it is time that dot maps graduated from high school and went to college. . . .
>
> PROFESSOR ROBINSON: I think they are in the kindergarten stage, and I think they ought to stay there. I would be very happy, indeed, if all dot maps that I have seen wouldn't have any legend at all on them. . . . Being scientists, geographers, planners and so on, you have a tendency to want data whether you need it or not.[45]

This exchange highlights the uneasiness, but the serious evaluation, of thematic mapping with computing machinery. The computer provided a much-needed interruption in the field, as decisions about surface representation could be rationalized, isolated, and brought into greater comparison.

Beyond the debates about map aesthetics and the role of data, this form of map production was a curiosity to cartographers. In a letter in 1966 between Robert Williams of the Yale Map Collection and Fisher, we can see how the method of interpolation is problematized. Williams writes, "In many cases the pattern obtained by the contour option of the SYMAP program is more a function of the program than it is of the data."[46] Williams included examples of SYMAP output to demonstrate the heavy-handedness of the program. Fisher would often annotate directly on correspondence, perhaps to help prepare his response. He circles and corrects the word *program,* stating that instead it is the "user's lack of knowledge" that causes such ill-derived patterns. Fisher understood that maps are imperfect representations. He continues in a response to Williams:

> I realize that this requires even more that the user of the mapping program understand what he [sic] is doing so that he will not be mislead—but this of course is true of practically everything in life—and certainly everything in the field of map making. As you pointed out this applies to map projections for example. If you try to judge the size of Greenland from a Mercator map you are certainly going to be badly mislead—and yet Mercator maps have real value for certain purposes. I don't think that we should deny ourselves certain values merely because a map is not perfect in all respects. No map is perfect. Every map is a compromise and all mapping depends on convention. The solution I think is to increase [peoples'] understanding of maps and how to read them.[47]

Williams also raised the issue of responsibility, given the persuasiveness of maps: "Since maps are perhaps the most persuasive form of communication, those who make them must accept an unusually high degree of responsibility for their truthfulness. This responsibility is increased when the aura of infallibility of the computer is added to the map."[48] Using his red pen, Fisher underlined *truthfulness,* drawing a line to the margins of the letter, and wrote: "all maps are a con." Fisher understood that the computer disrupted the relationship between the author of the map and the reader, a relationship that had always been based largely on deception.

SYMAP was originally developed on an IBM 709 with a deck of around three thousand cards. Amid many skeptics, Fisher felt a need to carefully innovate and demonstrate the significance of computing power for mapping. In a letter assuring a scholar at McGill, Fisher writes, "I understand fully your skepticism regarding the computer but that may well be because people tend to exaggerate what it can do. It is no substitute for human judgment or creative imagination, intuition, etc. But it is an invaluable tool that can be of vast aid."⁴⁹ Similarly to a member of the lab, Eric Teicholz, Fisher discusses the organization of a conference on the use of the computer in architecture. He writes, "In the computer field, when you start talking about what is possible within a decade, you immediately get into a realm of speculation that might not be very useful."⁵⁰ While he advocated against such hype, he undoubtedly believed that the lab was catalyzing an important and rapidly unfolding epoch. While considering a project team to do some mapping of Lynn, Massachusetts, Fisher chastises a young innovator of SYMAP who decided to stay in Tulsa, Oklahoma, and not take up his recruitment offer to move to Cambridge to work in the lab. "The world is moving," Fisher writes, "How much use are you making of the computer? What are you doing to move into the world of tomorrow?"⁵¹

To expose this moving world to SYMAP, a correspondence course enrolled more than five hundred participants in 1967. An innovation itself in terms of cartographic pedagogy, Fisher and his team designed a method by which computer mapping could be taught to the masses, with user instructions "to minimize the possibility of error and, through standardization, to facilitate the rapid checking of work."⁵² Participants would complete a coding form, using a SYMAP ruler, and then mail the completed coding form to the lab, where that form was used to create a series of computer punch cards. These would then be used on the computer to produce an output that would be mailed back to the participant. As a result, participants learned the basics of map techniques supported by SYMAP, and, perhaps more important, they learned "how one typically communicates with a computer, and something of the types of information that a computer requires to operate."⁵³ The dedication of the lab staff to ensuring this correspondence course's success would greatly expand the reach of digital mapping techniques.

After two years of freshmen seminars on problem solving, 1966–67 and 1967–68, Fisher proposed a new seminar, no doubt with encouragement from Bill Warntz, on spatial analysis with these computer techniques. The proposal outlined a range of disciplines that could be considered allied to this seminar:

> This seminar will deal with problems within the general field of spatial analysis, including but not limited to geographic space.... Within the broad field of spatial analysis, the computer has made it possible to do many things not previously feasible. Yet the use of the computer has demanded rigorous solutions for numerous problems previously unsolved, or solved only by unsatisfactory or intuitive methods.⁵⁴

However, Harvard College pushed back on such a seminar proposal, informing Fisher that it was too methodologically focused.[55] This undoubtedly frustrated Fisher, and likely agitated Warntz further.

The summer of 1969 included a trip to Swansea University, where Fisher met with a number of people engaged in computer mapping in the United Kingdom, including in London and in Edinburgh. Fisher reflected on the flurry of developments and clarified an agenda moving forward, toward a book on the mapping of quantitative information, of which he frequently spoke.[56] Computer mapping of quantitative information hinged on comprehension, made more complex in the move toward multiple subjects (multiple themes). During this summer, Fisher confronted a George M. Gaits of the UK Ministry of Housing and Local Government, who had set about creating a new mapping program called LINMAP inspired by SYMAP techniques.[57] The introduction of color into these processes, using a routine called COLMAP, likely vexed Fisher as this introduced new registers of experimentation. His systematic approach would require broadening.

Metadata

By 1970 Fisher was officially retired, maintaining a research title at Harvard without much salary. He was devoting much time to the book project, along with a growing interest in color models—with additional funding from the Ford Foundation to finish the writing. In this letter to Don Shepard in June 1970, his freshmen seminar student from 1966, he writes somberly:

> It appears that we were just a little ahead of our time. That's been the story of much of my life, and I don't recommend it. For your possible interest, I enclose a clipping with regard to General Houses, Inc. which I founded many years ago and of which I was the chief architect. My efforts in relation to computer mapping seem to have been better scheduled.[58]

Society seemed incredibly delicate; problems were abundant. Geoff Dutton, a student researcher under Warntz, recalls the setting of the lab in the basement of Memorial Hall:

> Olive drab canisters of crackers and drinking water were stacked in there, remnants of the fallout shelter craze of the late 1950's. This scene now seems a metaphor for how the Lab sheltered me from the fallout of academic politics and campus unrest as the sixties turned to the seventies.[59]

College campuses across the United States were erupting in protest. Harvard was no exception and many members of the lab were embroiled.

Fisher strongly believed that computer mapping was a solution of great potential, that with greater mapping would come greater data collection and better comprehension of global and local problems. In some passages that likely were preparatory for his book project, he clarifies:

> For—just as the purpose of computing is insight, not numbers—so the goal of mapping is understanding, not maps. To a very significant degree the problems of future war and peace may be favorably affected by the better knowledge to be achieved through the improved mapping of better data over coming years. It is, we understand, the position of various students of the Vietnam experience that, had there been a better understanding of the spatially variable facts in and surrounding Vietnam, the war would almost certainly never have been blundered into by either side.[60]

Had more information been made more comprehensible, he believed "we probably would not have gotten into the war."[61] Indeed, Fisher had intervened with at least one lab member, advocating for his Harvard appointment to appease the draft board.[62] Fisher continued to connect the Vietnam War with urban management:

> To the extent we are necessarily concerned today with difficulties to be found in our large and rapidly growing cities, not only in the United States but in most countries of the world, the problem is of a similar nature. The difficulties of our ghettos, for example, involve as a necessary and inherent fact spatially variable information in and about the ghetto . . . For an adequate understanding of ghetto problems in relation to the city as a whole . . . it is clear that an adequate knowledge of the spatially variability of pertinent information is crucial. As in the case of Vietnam, the facts of value that are available today far exceed the facts which are being used. In policy making and in other connections, we are not employing effectively more than a fraction of the information that is available to us—and we are not sufficiently encouraging the collection of still better and more valuable data as needed.[63]

While the Fisher papers do not clearly demonstrate nor explain the involvement of the lab in the war effort, correspondence between Fisher and Tom Thayer, the deputy director of intelligence and force effectiveness of the Office of the Assistant Secretary of Defense, in March 1967 details the needs of the National Military Command Center for a "graphic display system" to analyze data sources to include friendly and enemy forces, enemy incidents, and friendly casualties, toward the "integrated analysis of the situation in Southeast Asia."[64] The story of the lab's specific role in the computer mapping of the Vietnam War has yet to be written.[65]

In March 1970 Fisher facilitated a major conference on the mapping of quantitative information in Evanston, Illinois, at Northwestern. The event was chaired by four leaders in the cartographic establishment—George Jenks, Arthur Robinson, John Sherman, and Waldo Tobler—with whom Fisher had much one-sided correspondence. Tensions seemed palpable. From the transcripts of the conference, Jenks appeared annoyed with Fisher's contributions, as well as the contributions of members of the lab. Computer-mapping approaches attempted, argued Jenks, to have their cake and eat it, too. Jenks continued:

> You cannot have the capabilities of all of the data and at the opposite end have the capabilities of the graphic image which summarizes and generalizes a concept at the same time. . . . I maintain that the tradeoff is that you have ruined the generalization in order to save the data.[66]

Generalization and preservation of data were sore topics. Fisher was quick to raise issues of readability and comprehension, and he encouraged Robinson to intervene on aesthetics, without much effect. For Fisher the computer made much of traditional thematic cartography inappropriate to the task at hand.

Indeed, for Fisher, "The computer has forced new thinking."[67] On this subject, he was unwavering. To close the loop, he attempted many times to enroll Horwood at the University of Washington, back into the project. In a memo to Carl Steinitz in 1967, Howard encouraged working with Daniel Conway, a student of Carl Rogers from the University of Pittsburg, who was a student of Horwood at Washington.[68] These lineages mattered, in a time of sweeping change. However, in his later years, Fisher wanted to establish where Horwood left off and SYMAP began. In 1975 he wrote to Horwood to recover that particular history, "In a word while the SYMAP program was not an outgrowth of your program—in the sense that it was not built upon it—it was entirely the result of the stimulating experience which you provided to me."[69] Later in 1975, in a memo to Allan Schmidt, he clarifies:

> There is no question that Ed's course led to its development—but as a form of rebellion against his so-called maps which were merely numbers printed on plain white paper. . . . Ed never did produce by computer anything that could be called a map—at least previous to the development of SYMAP.[70]

In 1975 and 1976 much of Fisher's correspondence sought to thank those involved in the earliest days of the lab (including Betty Benson, Waldo Tobler, and Brian Berry). His own reconciliation with the history of SYMAP with Horwood found its way into the introductory materials of SYMAP.

The lab was in peril during much of the 1970s. Bill Warntz resigned in a huff in 1971, writing to Fisher that "the conditions within the Graduate School of Design . . . represented to me the most self-defeating set of arrangements I had ever observed in 22 years of active research and teaching."[71] Indeed, the GSD considered closing the lab in 1974, which coincided with the death of Bob Weinberg. Weinberg was a GSD alumni and friend of Fisher who had helped support the formation of the lab before Fisher won the Ford Foundation grant. Fisher quickly negotiated a donation from the Weinberg trust to the GSD as a shot in the arm. A Computer Graphics Prize was quickly established, and Brian Berry, another Garrison geography grad, was parachuted in from the University of Chicago to direct the lab. By 1978 software from the lab was in operation in twenty-six countries outside the United States. The lab was experimenting with new techniques and new software, with rapidly evolving computing machinery and display technologies.[72]

Howard Fisher died in January 1979 at the age of seventy-five, leaving much of his work on color models unfinished, as well as the book project funded by Ford. His friends and colleagues set about producing a text based on his manuscripts, led by Allan Schmidt, and a local publisher was found to release the book in 1982: *Mapping Information: The Graphic Display of Quantitative Information*. Berry provided a nice summation of Fisher's effect in the foreword:

> He was not shy about challenging the sacred cows of traditional cartography, much to the discomfort of many who preferred tribal incantations to the systematic principles of map symbolism that Howard felt to be essential.[73]

The book emerged in a field that was rapidly changing. It offered little in the way of the history of computer mapping, and instead focused on Fisher's experimentation with classes and generalization. Few seemed to pay attention. Robinson's fifth edition of *Elements of Cartography* would emerge in 1984, largely unscathed by the lab's work.

Fisher recognized the power of maps long before cartography would turn critical. In the opening of his book, he wrote, "There is a genuine need for a small book entitled *How to Lie with Maps*."[74] Indeed, Mark Monmonier offered one of the only reviews of the book in 1984.[75] While the book is a bit of a time capsule of midcentury, wide-eyed thought in computer mapping and its impacts on thematic mapping, it is these asides—like a call for *How to Lie with Maps*—that give me pause. Mapping, then as now, was more than a tool. For Fisher, map applicability was deeply connected to comprehension, even psychology. This went back to his earlier conversations with Williams at Yale:

> All maps are abstract and approximate in one way or another as Bill Warntz tried to bring out. The question is where do you draw the line and refuse to make a map at all for fear it might confuse someone unqualified to read it?[76]

And as we return to and reconsider the digital methods developed at the lab, that experimental spirit remains—that the line of comprehension should not limit mapmakers, but should instead inspire them toward new ends. This is perhaps Harvard's most important contribution to the practice of twentieth-century geography—a contribution born precisely in the context of a failed academic discipline at Harvard. The year 1948, therefore, is not only a moment of continental disappointment but perhaps a moment that also "cleared the deck" for something else, something new, quite vulnerable and risky, that might actually change the game.

CHAPTER THREE

Movement
Strange Concepts and the Essentially Subjective

> Now, what do I mean by time. First, there is nothing of value or tangible about time per se. The concept of time is extremely useful only because it allows us to study processes effectively—processes which require inputs and yield outputs, where some of the inputs and outputs are of value, although sometimes they are of intangible character only. Viewed another way, our interest is in the innumerable social and physical processes of actuality and their interrelationships.
> —WALTER ISARD, "ON NOTIONS AND MODELS OF TIME"

> Cinema is not a universal or primitive language system [*langue*], nor a language [*langage*]. It brings to light an intelligible content which is like a presupposition, a condition, a necessary correlate through which language constructs its own "objects" (signifying units and operations).
> —GILLES DELEUZE, *CINEMA 2: THE TIME-IMAGE*

Time is an elusive concept—nearly as difficult as space. Time announces itself, it would seem, each day, each night. Planetary movements are represented by the minutest changes on the faces of our clocks and watches. Walter Isard, one of the primary founders of the field of regional science, worked at MIT down the street from the Laboratory for Computer Graphics and Spatial Analysis at Harvard. Many of Isard's papers were shared with members of the lab by Bill Warntz. (Both Warntz and Isard were active members of the Regional Science Association.) Isard understood the explicit utility of time, while also recognizing that there were myriad other intangible qualities. To read Gilles Deleuze on cinema alongside Isard is to take up these intangibles: the force of such a concept of time, their "actuality and their interrelationships," quoting Isard, "brings to light," quoting Deleuze, in ways that capture our attention.[1]

The rise of GIScience can be understood, from one perspective, as an elaboration on centuries of practice to capture time, to render time immobile, in order to fix(ate on) spatiality. As I suggested earlier, the role of the map has been to slow the speed of time, to grant a pause such that we might consider the moment.[2] Isard recognized this as a key aspect of the study of social and physical processes—indeed, computational spatial analysis inherits these thoughts. It is now commonplace to hear scholars of GIScience bemoan the treatment of time in spatial software. Certainly, the sins against time are confessed nearly as often as the sins against ethics in GIScience pedagogy and research: just as "ethics" becomes a unit in the *Body of Knowledge,* so, too, does time.[3]

Enter big data. These kinds of representations have reached the category of genre—maps representing the rhythms of bike share programs in London or New York or maps of globally trending Twitter hashtags are reproduced in scholarly journals, online news media, and media aggregators. Movements across these big data maps produce a fine gloss, washing over the map reader–viewer, slack-jawed in a cinematic embrace. The rapidity of these datasets complements nicely their proliferation and ubiquity. Time appears to speed up through these spatial representations. As presuppositions, these maps hint at relationships and conditions and call our attention to the surface. Swirling patterns and clusters allow us to point fingers and place blame, as well as shower rewards and celebrate idiosyncrasies. Time and space seem sutured in these quick representations; they resist simple categorization as spatiotemporal. These are innumerable. In ways I discuss below, big data maps are cinema. They bring to light.

I take up the historical precedents of this genre of big data maps, reviving a critical impulse of qualitative GIS: movement. Here, I suggest movement as an interesting foil for the rigidity of efforts to claim territory around spatiotemporal models in GIScience. Movement, to move and be moved, is both what the new lines of our maps attempt to produce for the map reader and what our maps attempt to document. I focus on a concept of maps that move to think through an engaged digital mapping that calls into question the systemization and capturing of knowledge-power by intervening in widely held practices of spatial representation. I explore the implications of digital mapping on knowledge production, including the role of the objective and subjective, representation and the more-than-representational, the politicization of knowledge, and the stylization of data. I begin by reengaging the map communication model as it emerges from the writings and practices of Arthur Robinson, by unpacking the map percipient in the context of interactive and dynamic cartographies, where the map reader is increasingly a data product. I trace this through a selective history of animated cartography, from the work of Waldo Tobler, Allan Schmidt, and Geoff Dutton, introduced in chapter 2. In doing so, I insist on an evolving (and nagging) question at the center of map design: How might mapmakers move beyond design thought as cartographic information efficacy toward design thought as worlding intervention?

In previous work I have been invited to think about the possibilities of qualitative GIS and the envisioning of a different practice of mapping.[4] For me, the critical impulse of qualitative GIS was in its role as *dispositif*, as the kind of machinery or apparatus to cause power to move, to become capillary.[5] The idea of qualitative GIS was such a kind of grotesque discourse for me, serving to rearrange and rethink the representational systems at the heart of GIScience; indeed, the idea of qualitative GIS was destratifying. However, I feel its deployment has fallen short on this potential, as many invocations of qualitative GIS are perhaps better thought as new systems for capturing, analyzing, and representing data that happen to be nonquantitative: restratification along the line of flight.

Instead then I want to try again and sidestep somewhat the idea of qualitative GIS, to more directly take up maps that move as a way to question the role of systemization and capture in geographic representation. And in doing so, I want to take up these particular maps that move—animated maps. The point of this work is not to exhaustively articulate the individual techniques of animated cartography, nor provide a comprehensive listing of animated maps, but rather to discuss those lines that make movement possible within maps. This is central to what has been called a Deleuzian empiricism, as that which

> is not interested in language as such but rather in what makes it possible, what confrontations, what relations of forces, what viewpoints, what events are folded in, enveloped in signs.[6]

This is where Deleuze's writings on cinema have tangency with the project on the drawn and traced lines of our maps. *Cinema* sought to understand the Bergsonian thesis on the materialization of memory, not by casting aside cinematographic technologies as false memory[7] but to explore the variety of ways in which cinema (its technologies, techniques, and traditions) illuminates the openness of the universe.

Cinema, as the moving image, provided a useful touchpoint for understanding the richness of thought, action, and force, the various comings

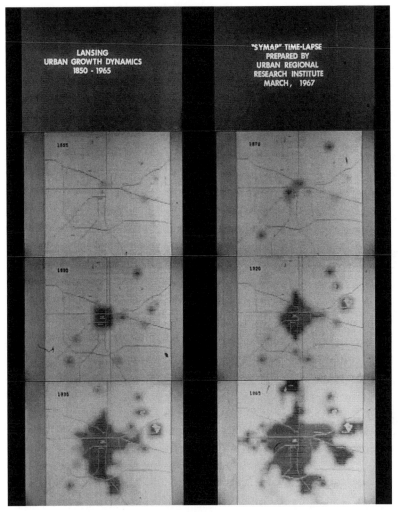

FIGURE 15. Stills from Allan Schmidt's 1967 animated map of Lansing, Michigan. Courtesy of Allan Schmidt.

together and contingent departures. Throughout the text, it becomes clear that Deleuze is not interested in producing a theory of film, but in understanding a metaphysics of movement and image:

> It is not the line which unites into a whole, but the one which connects or links up the heterogeneous elements, while keeping them heterogeneous. The line of the universe links up the back rooms to the street,

the street to the lake, the mountain, the forest. . . . Each one of us has his own line of the universe to discover, but it is only discovered through tracing it, tracing its wrinkled stroke.[8]

The punctuated line of the everyday is understood only through tracing the connections that may appear at first as whole compositions. For me, the animated map occupies this space of a *determinant indeterminacy*, as the mush of the moving map exceeds the combination of lines drawn in composed snapshots. In other words, the moving map is a product of great technical and creative facility, and yet, when traced, it produces so much more than individually drawn snapshots would reveal.

Therefore, when considering the variety of innovations in animated mapmaking, I find it helpful to begin from continuities rather than draw neat distinctions. In 1967, after Allan Schmidt had moved to work at the Harvard lab, his team back at Michigan State took to creating an animated map of Lansing, Michigan. Figure 15 captures a series of stills from this animation on 16mm film.[9] The film was created from timed exposures to static map outputs from SYMAP, creating the illusion of an urban morphology actively expanding for the viewer. Compare the animated map of Twitter posts related to the riots in Ferguson, Missouri, in 2014, stills shown in Figure 16, created by CARTO and appearing in the *Washington Post*.[10] As the timescale advances, this animated GIF map (of 100 frames) seems to represent an eruption of activity at major population centers, paralleling the passions and intensity of the event. Both of these animations are born out of incredible leaps in computing and geographic representation, although separated by nearly fifty years. Both attempt to represent the spatiality of social phenomena across a map surface. Both are meant to inspire as well as instruct. And yet there was and is certain uneasiness with these productions, as they rub against an establishment cartography as well as a persistent critique of the command and control of geographic representation.

Far from attempting to resolve these tensions and uneasiness, I hope to get a little muddy in their mixing. In what follows I ruminate with some map animations from the map archives, while reading them with and against the grain of cartographic scholarship and spatial theory. It is in the drawing of these lines that the techniques of cartographic animation and the techniques of cinema are sutured—enabling ways of thinking and acting through map animation beyond the purely effective. Then as now, experiments with maps that move disrupt simple attempts to declare the objective in the cartographic. These animated maps and the affective response we give them disclose different capacities, which have perhaps always

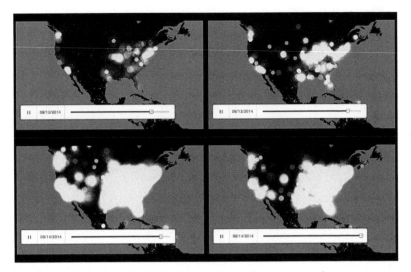

FIGURE 16. Stills from CARTO's animated map of user-generated posts on Twitter related to the Ferguson, Missouri, protests August 13–14, 2014. Map created by Simon Rogers, reproduced by permission of CARTO.

exceeded the neat lines that bound our maps. Animated maps reveal potential epistemological irreducibilities in the tradition of cartography. The point here is not to reconcile their remainders, but, as Haraway insists, to "stay with the trouble," to trace consistencies and continuities while highlighting breaks, or punctuations, in the line.

The Look of MyMaps®

Arthur Robinson's (1915–2004) dissertation at Ohio State, which he defended in 1947, was the proving ground for two publications that contribute to his veritable legacy: *The Look of Maps* (1952) and *Elements of Cartography* (1953).[11] At the moment of their writing, Erwin Raisz's *General Cartography* text was arguably the dominant voice in cartographic pedagogy.[12] Under the breath of the second edition of this text in 1948, cartographic practice was beginning to more explicitly wear the garments of science and the techniques of war.[13] But these shifts in geographic representation were happening quickly, requiring more than a revision of existing textbooks. At the center of this rapidly changing domain was a growing recognition of maps as a visual *system* of correspondence between map reader, the map, and the mapped reality. Robinson would suggest (and later become more insistent) that "there is probably room for argument . . . that the "art" in cartography should be considerably more objective than it has been in the past."[14] Indeed, his feeling that "functional design" should be central to the creative process in mapmaking was considered a significant attitude adjustment in the field.[15]

The target of Robinson's critique was not specific cartographic conventions, per se, but rather the process of innovation itself in cartographic practice. That "convention has merely replaced convention" highlighted an unaddressed ugliness in the tradition. He continues, "All too often, however, they have been tested only by their makers, and the quality of parental pride is not always objective."[16] At the heart of this critique was Robinson's displeasure with statements from established cartographers that effective map techniques (understood as the correspondence between the map and the phenomena mapped) required artistry that was largely subjective. Too often, he argued, a map's ineffectiveness was merely attributed to the eye of the beholder. Beauty, pleasure, and joy in mapmaking and map reading were distractions from data, communication, and the general and dispassionate assessment of the effectiveness of the map.

Just as mapmaking would undergo rapid changes in the postwar period, the solidification of a science of cartography would enable a new generation of mapmakers concerned with the measurement and fine-tuning of the correspondence between the map and the mapped. Experimentation with mapmaking conventions would proceed scientifically, and thereby objectively, untethered from the inheritance of techniques and artistic style. The more subjective aspects of cartography, previously based on "convention, whim, and fancy," could be isolated and analyzed as mere "characteristics of perception," subtended to the distinct elements of cartography.[17]

The fervor of this approach is what I would summarize as the specifically Robinsonian tradition. From these moments of the 1950s would grow a new set of adherences, beyond conventions replacing conventions, toward an exacting and reasoned development of technique. It is within this Robinsonian refashioning of cartography that digital mapmaking methods were tested.[18] While Robinson himself seemed initially suspect of the incursion of new, computationally derived lines on maps, his approach could adapt and incorporate these techniques, given the appropriate study and evaluation of their impacts on perception. It is important to note that Robinson leans heavily on the debates and discussion within the field of advertising, as the domain of the science of perception of visual compositions.[19] Indeed, advertising is leveraged as the state of the art in the deliberate and evaluated incorporation of new graphic techniques. And as the significance of this will become clearer in chapter 4, the links to advertising and marketing in the Robinsonian tradition is pronounced in mapmaking engagements in the contemporary attention economy.

Robinson's cartographic largess was underscored by his commitment to historical studies of mapmaking.[20] However, in the era of MyMaps® such attention to dusty maps would be limited. As maps became the explicit tools of advertising and marketing, knowledge of what came before is devalued in favor of the arms race in establishing the primary platform for a particular and profitable location-aware future. MyMaps®, a reference to Google My Maps, is my shorthand for the contemporary technologies, industries, and scholarship of interactive, web-based map design. The look of MyMaps®, then, is an elaboration of the Robinsonian drumbeat of functional design and the studied development of visual compositions as part of a broader system of user-map-reality. This neo-Robinsonian approach, then, attempts to capitalize on the great lengths to which cartography adopted principles of map function, such as visual variables and later, interaction primitives.[21] Neo-Robinsonians adhere to a progressive understanding of the map, a tight relation between deliberate adjustments in map interface and a corresponding improvement in user experience. No longer bound by the rhythms of scholarship, this sentiment in map design is nestled securely within the software and web design community.

However, despite the neo-Robinsonian necessity to evaluate and strengthen the objective elements of (carto)graphic practice, Arthur Robinson acquiesced on some fronts, which is of importance to this discussion:

> Many of the aspects of harmony, movement, balance, and proportion, seem likely forever to remain essentially subjective insofar as their evaluation is concerned. This does not mean to imply that the principles governing their use are purely a matter of individual caprice; it does mean that *exact* standards probably cannot be devised.[22]

That Robinson opens the door to harmony, movement, balance, and proportion, as those ineffable, "essentially subjective" qualities of geographic representation, underlines the simultaneous attraction and elusiveness of the animated map.

Spatiality and Animated Ectoplasm

It may come as little surprise that much of what Robinson declares to be the resolutely subjective elements of cartography are perhaps among the livelier indicators of spatiality in critical geography. Movement, balance, harmony, and a sense of proportion exceed the drawn lines fixed on the page and screen. These four, but specifically movement, conjure a subject—not a percipient in a neo-Robinsonian map communication model,[23] but the expression of a directionality of force. These four presuppose the percipient; they are presuppositions of the percipient. However, this does not stop more resolved cartographers from attempting to pin down these more slippery elements. Herein lies the trouble.

When considering the contemporary relevance of such problematization, I find it instructive to think about the rub between two interventions and provocateurs, however unlikely they may be grouped together: the critical human geographer, Doreen Massey, and Kenneth Field, a cartographer currently working at Esri. Massey's ruminations on space, time, and representation beg geographers to unsettle our most basic assumptions and consumptions regarding spatiality, liveliness, and the implications of spatial representation. In a different register, Field brings the weight of the tradition of cartography (its conventions and priorities) on the proliferation of maps, mapping kitsch and clichés, and, more specifically, big data maps. Each of these figures engages in map critique; each articulates a series of shortfalls in geographic representation. Each helps us recognize the presuppositions of perception, although with different object lessons.

In *For Space*, Massey gives us a kind of spatiality hymnal, tracing the difficult but worthy path of discovery: spatiality is never quite what it seems. Much of the text navigates between poststructural philosophers and attempts to constitute the basic furniture at the heart of a concept of space. What bothers Massey is summarized at the beginning of the text:

> Equations of representation with spatialisation have troubled me; associations of space with synchrony exasperated me; persistent assumptions of space as the opposite of time have kept me thinking; analyses that remained within the discursive have just not been positive enough.[24]

As a critical human geographer, Massey underlines a persistent worry of the discipline—that an imagination of space is largely limited to the dominant forms of representation available. Geographers' work to understand spatiality is largely constrained to a series of representational tricks that ultimately limits their range of movement. The progressions of relations and contradictions that produce what is all too conveniently defined as globalization, or the evocations around place that presume too much and yet say too little. She continues to document the implications of an overly simple equation of representation with space:

Representation is seen to take on aspects of spatial*isation* in the latter's action of setting things down side by side; of laying them out as a discrete simultaneity. But representation is also in this argument understood as fixing things, taking the time out of them. The equation of spatialisation with the production of "space" thus lends to space not only the character of a discrete multiplicity but also the characteristic of stasis.[25]

That space becomes characterized as stasis through representation is the specific difficulty confronted by geographers, as it belies the ways in which spatiality operates. From this perspective, it is not enough to recognize space as a production (a leap for many who adopt a more conventional notion of space).

As she works through these implications, Massey attempts to sort a series of formulations offered a century earlier in *Matter and Memory* by Henri Bergson and revisited by Gilles Deleuze.[26] Bergson questions an assumed opposition between duration and space, and in the words of Massey, he attempts to think through the "duration in external things and thus the interpenetration, though not the equivalence, of space and time."[27] The trouble of the relationship between duration (as time) and spatialized representation (as space) is also the rub of Massey's argument. Here, she advances a notion of space:

> as the dimension of multiple trajectories, a simultaneity of stories-so-far. Space as the dimension of a multiplicity of durations. The problem has been that the old chain of meaning—space-representation-stasis—continues to wield its power. The legacy lingers on.[28]

And this is perhaps where our slack-jawed watching of animated maps begins to take a critical turn. There is an incredible power or force, not fully understood by the neo-Robinsonian drive to reduce the map to a series of perceptual relations, that draws our attention to the animated map. Massey continues:

> Space conquers time by being set up as the *representation of* history/life/the real world. On this reading space is an order imposed upon the inherent life of the real. (Spatial) order obliterates (temporal) dislocation. Spatial immobility quietens temporal becoming. It is, though, the most dismal of pyrrhic victories. For in the very moment of its conquering triumph "space" is reduced to stasis. The very life, and certainly the politics, are taken out of it.[29]

To the chain user-map-reality of the neo-Robinsonian tradition, we can add then the chain space-representation-stasis. These chains of significance are undoubtedly entangled, as the latter does not necessarily emerge from the former but is certainly strengthened by it. In other words, what Massey identifies as space-representation-stasis is a chain of significance that reaches deep into the representational tradition of geography, while user-map-reality is but a specific, systematic intensification of that tradition.

Animated cartography, those pesky maps that move, further test these chains of contemporary representational practice. They attempt to capture the life of space: as spatiality. Do they not attempt to resolve the inadequacies of the static map, where time is merely a condition? Certainly, these maps move: their lines unhinge and our desires are stirred. However, can these maps resist the conquering of time and the reduction of space to stasis? Here, I introduce another reaction to these types of representations, although less philosophically sublime. The writings of behavioral cartographers on the topic of animated maps gather similar energy, while performing different critique. At its most abbreviated, fewer than 140 characters, is the provoking tweet by Ken Field in 2014, "I'm wondering when people will realize the animated ectoplasm twitter maps don't actually show anything."[30] Undoubtedly a lob at CARTO and the proliferation of these kinds of animated maps of Twitter updates, Field's critique might also be read alongside a persistent uneasiness in the development of animated cartography. As an employee of Esri, Field may be channeling a particular skepticism of the crowded web-mapping industry and the attendant prioritization of the gloss of an interactive web map over the ability to tell a specific story, to take a position.

For Field, animated ectoplasms extend the chain of space-representation-stasis, and while there is much to his critique, we might draw parallels to Danny Dorling's suggestion more than twenty years earlier, regarding animated cartography:

> There is no reason why the map should remain fixed while the action is played out upon it. Indeed, there need not always be a traditional map in every frame of an essentially cartographic animation. The barriers that separate us from the other visual arts may disappear as we begin to use the same machines. We need to learn from the experiences of film and documentary makers and from the computer games market, and to use their tools (which often are easily accessible).[31]

Similar to Field, Dorling is exasperated by the conventions of animated maps, that despite their ability to capture our attention, they do very little to communicate the complexity of spatiality or even, as Field suggests, the subject they ostensibly are meant to represent. That Dorling would suggest film as a domain that cartographers should investigate is instructive of the limits of the cartographic model as fashioned by the neo-Robinsonian imagination, just as it also extends the neo-Robinsonian impulse to look afield toward those domains that capture human imagination and the methods and techniques they successfully employ. In the 1950s it was advertising for Robinson; by the 1990s it was film and gaming for Dorling.

While Massey does not directly address animated maps, we can perhaps read into her critique of the stultifying effect of these geographic representations on space. Here, cartographer and spatial theorist alike are unsettled by the animated map, perhaps sharing in a recognition that in the worst case these representations serve to stabilize and depoliticize space and spatial relations, and at best they are just not effective devices for communicating information about spatiality.

But let us stay with the trouble. Read through the lens of neo-Robinsonian cartography, it is the possibility for maps that move to operate *affectively*—that is, beyond those elements of cartography that can be measured for their effectiveness—that requires new conceptualization. That space of "multiple trajectories" and of "stories-so-far," which activates Massey's critique, might be found in particular maps that move, just as other maps that move, those animated ectoplasms that vex Field might also foreclose the possibility of achieving that effective moment, of a transference of understanding in the correspondence between reality and map. At this point, then, I suggest we need further conceptual development to take up those more subjective elements of cartography, which cause map readers to gaze unendingly at the moving spatial representation in front of them.

I introduce another theorist into the mix, to aid my thinking the implications of animated maps as cinematic devices. Deleuze begins *Cinema 1* by working through Bergson's thesis from *Matter and Memory,* by bearing out and elaborating this thesis about memory, movement, and image using twentieth-century cinema (which Bergson brushes aside as a kind of false movement). Deleuze writes, to clarify this Bergsonian position, "In short, cinema does not give us an image to which movement is added, it immediately gives us a movement-image."[32] The operations of cinema to produce a movement-image, a concept from Bergson, is important to our discussion of animated maps as technosocial artifacts. Deleuze continues, discussing the technical conditions of cinematic production, to include:

> not merely the photo, but the snapshot . . . the equidistance of snapshots; the transfer of this equidistance on to a framework which constitutes the "film" . . . a mechanism for moving on images . . . It is

in this sense that the cinema is the system which reproduces movement . . . as a function of equidistant instants, selected so as to create an impression of continuity.³³

In this brief passage, Deleuze takes up the technical specificity of cinema at the level of the film: the perforation of the film and the mechanical means by which the film is advanced. These equidistant instants, technically speaking, are effective because they become assembled in human perception as movement. In cinema, suggests Deleuze, "any-instant-whatever" is the equidistant instant:

> We can therefore define the cinema as the system which reproduces movement by relating it to the any-instant-whatever. But it is here that the difficulty arises. What is the interest of such a system? From the point of view of science, it is very slight. For the scientific revolution was one of analysis. . . . Did it at least have artistic interest? This did not seem likely either, since art seemed to uphold the claims of a higher synthesis of movement, and to remain linked to the poses and forms that science had rejected. We have reached the very heart of cinema's ambiguous position as "industrial art": it was neither an art nor a science.³⁴

I suggest that Deleuze's articulation of the "problem" presented by cinema draws to a finer point that which Robinson describes as the limits of standardization in some elements of functional, cartographic design. That movement is too difficult an element to render objective and measured might be understood otherwise as the difficulty of animated maps as a kind of "industrial art," neither built for the rigor of scientific analysis or the interpreted subtleties of form.

In *Cinema 1*, Deleuze interrogates what he terms the movement-image, of which he describes three varieties of film techniques: the perception-image, the affection-image, and the action-image. That this text was not to be a theory of film, but a conceptualization of the image, allows some productive borrowing of this philosophy in our own understanding and taxonomy of the betwixt position occupied by maps that move. And while there is much to be unpacked in the application of the movement-image concept on animated maps, I highlight the role of the specific capacity of the affection-image and its relative taboo in cartographic practice.³⁵ The affection-image, according Deleuze, is demonstrated in film by the close-up and the *expressive face* (pure power, understood as the affection-image that carries us from one quality to another) or *reflective face* (pure quality, understood as the affection-image that proposes a commonality among many objects).³⁶ Already we have two aspects of the potential affectivity of the animated map, which further opens those more subjective elements of cartography.

Neither Art/Science

The development of animated maps mimics an ambiguity described by Deleuze, of cinema as neither art nor science. However, while there have been significant attempts to assert the science of animated cartography, it has been a slippery object. Deleuze's discussion of use of the close-up allows us to better understand the technique of the affection-image and the ways in which these movement-images are always more than their industrial components. In drawing on film techniques from Georg Pabst's *Pandora's Box* (1929), he writes:

> There are Lulu, the lamp, the bread-knife, Jack the Ripper: people who are assumed to be real with individual characters and social roles, objects with uses, real connections between these objects and these people—in short, a whole actual state of things. But there are also the brightness of the light on the knife, the blade of the knife under the light, Jack's terror and resignation, Lulu's compassionate look.[37]

In the play of objects and subjects, there are affection-images that express both a change in quality (from passionate intimacy to intense terror in the face of Jack) and a correspondence among Jack's face and the movement of the lamp's light along the blade of the knife. As types of movement-images, these moments in the film enable Deleuze to move beyond the "actual state of things" found in the composition of the frame itself, toward a virtuality of things—becomings. I suggest that expressive and reflective faces, as two poles in an affection-image, can be used to identify similar taxonomies in maps that move.

This brief and explicit moment from *Pandora's Box* represents how the close-up creates intensities and shifts the qualities from object to individual to object, and so on. The composition of the frame, the individuals within the frame, their stories and characters, come together to create what seems at first as a closed system: a set. However, the flash of the knife, despite its composition in frame, is already "out-of-field." Deleuze understands this as indicating two potentially interrelated aspects—out-of-field as a relation to something elsewhere or out-of-field as "a more disturbing presence, one which cannot even be said to exist, but rather to 'insist' or 'subsist', a more radical Elsewhere . . . the adding of space to space."[38] The moving light on the knife intensifies in the eyes of Jack, transferring a specific quality from object to individual to object through a series of close-ups. Neither science nor art. But consider the advance of the animated map.

Arthur Robinson and some of his students set about working through the opportunities that animation would bring to cartography, alongside broader changes with automation in the field in the 1950s. One such student, Norman Thrower, establishes the key furniture in the development of the field, in a piece titled "Animated Cartography" published in 1959. He writes:

> Although in the past the animated drawing has been associated particularly with entertainment and advertising, it is being used increasingly for scientific illustration. Maps lend themselves particularly well to animation; by the use of this technique we can add another dimension to cartography—time.[39]

Thrower situates the emergence of animation in scientific graphics within the National Defense Education Act of 1958, which funded the training and use of new visual technologies. This injection of cash was, of course, a response to concerns that education in the United States was falling behind the Soviet Union, following their launch of Sputnik.[40] The leap toward machinery and training that would enable new representation of scientific approaches would be part of this geopolitical strategy. Thrower, drawing similarly on Robinson's interest in advertising, attempts to reclaim the "animated drawing" from the clutches of entertainment, to add a temporal dimensionality to cartographic representation.

Thrower overviews the technique of producing animated cartography at midcentury. Maps are created and traced on celluloid (cels) that are framed, registered, and fixed below a mounted camera. Thrower continues to describe the parameters of the creative process:

> It is sometimes desirable to illustrate two different, simultaneous movements; for example, the expansion of a city area and population growth. This can be accomplished by using two cels for each exposure. . . . Visual "punctuation" of the film can be achieved by such techniques as the "cut," "mix," and "fade."[41]

To the studied techniques of static maps, time as movement was added. This introduces new configurations, new possibilities. Cut, mix, and fade became techniques of style, the cartographic equivalent of the cartouche.

By 1966 Bruce Cornwell and Arthur Robinson would review major innovations in the field of animated cartography. A number of computing developments including the CRT display and drawing "light pen" further evolved the process of successive frame animations with film, drawing

FIGURE 17. Edward Zajac's 1963 "Simulation of a Two-Giro Gravity Attitude Control System" was reviewed by Cornwell and Robinson as an innovation for animated cartography.

directly from the film industry. Figure 17 displays work by Edward Zajac at Bell Labs in 1963, reviewed by Cornwell and Robinson. They argue for the potential of this technique, specifically in combination with the computer: "Certainly the prospect for the use of this versatile technique in accomplishing creative dynamic cartography is limited only by the imagination."[42] The creation of vectors that could be quickly rendered on-screen and transferred to a recording enabled new ways to represent spatial movement.

But this creative process would prove challenging for a cartographic establishment made more scientific through the careful assessment of the quantification of perception. Jacques Bertin would insist in 1967 (although not translated into English for more than a decade) that movement was an "overwhelming" variable. He writes that movement

> so dominates perception that it severely limits the attention which can be given to the meaning of the other variables. . . . real time is not quantitative; it is "elastic." The temporal unit seems to lengthen during immobility and contract during activity, though we are not yet able to determine all the factors of this variation.[43]

Bertin, who would become well known for an exhaustive delineation of the visual variables of cartography, attempts to document the parameters of any graphic representation of movement. And while these are instructive of the way in which the midcentury science of cartography would be established and secured, more interesting is his recognition of the indeterminate elasticity of time in spatial representation. Regardless of the inability of thematic cartographers to determine the factors governing movement in map design, animated cartography was the subject of much experimentation in the earliest days of the digital map.

As already introduced previously and in Figure 15, "A Pictorial History of the Expansion of the Metropolitan Area" of Lansing, Michigan, was created by Allan Schmidt in 1967, as he began to make plans for his departure from Michigan State to Harvard. At MSU he adapted the digital mapping program called SYMAP to create a series of static maps of the development and expansion of the city. These were then taken to a visual graphics lab at MSU, where a 16mm film was created to capture the movement-image of an urban morphology. While the film quality attests to the day in which it was created, the intensity in which this close-up of the "face" of the graphic representation of the city is meant to create a sense of the transformative rhythm of progress. The three-and-a-half-minute film has three sections, each an experiment with the speed of the exposures, where a SYMAP static map in five-year intervals (1850 to 1965) was recorded in increasingly shorter exposure times. This temporal dimension was no doubt a playful parameter, as Schmidt could control the rhythm of the growth of the urban morphology. The elasticity of time produces a fine gloss, where the gradual expansion of the city races toward continuity, mirroring a sense of the naturalness of urban development.

Waldo Tobler and his student Frank Rens at Michigan were similarly experimenting with these digital methods, including what would eventually be called SYMVU, which allowed for the drawing of three-dimensional surfaces.[44] In these experiments, Tobler, like Schmidt, also adjusted the intervals of exposure to these successive map snapshots. However, the oblique perspective on the city further emphasizes the naturalness of urban expansion and the idealized elasticity of time. It is in these experiments from which the so-called Tobler's first law of geography emerges: "everything is related to everything else, but near things are more related than distant things."[45] Space, fixed by these SYMVU outputs, was made to move through the repetition of these "equidistant instants," using Deleuze. Tobler attempted to fine-tune these qualities, and his development of a most basic theory of absolute relations in space ironically emerges precisely from experiments with representation of spatial movement— one of the more difficult elements of cartography to generate objective measures.

By the late 1970s these creative experiments in animated cartography were bountiful. Many of these representations were clearly about the affectual dimension of maps. Hal Moellering, another student of Tobler and John Nystuen, summarized his work with filmed instants in a project to map traffic accidents in the region around Ann Arbor, Michigan: the point was to "develop a feel for a spatial and temporal dynamics associated with traffic crash production."[46] Moellering would go on to explore real-time animations with 3-D maps, enabling the user to navigate and visualize spatial processes.[47] The "feel" of space was certainly seen to be lacking from digital static cartography of the day. Animation would challenge immobile representations of space by insisting "the adding of space to space," in Deleuze's words. However, these more radical presuppositions advanced by the moving map were sources of trouble. With its roots firmly in the quantitative traditions born out of the University of Washington and the thematic cartography traditions of the University of Wisconsin, experiments in animated maps attempted to tame and define the ineffable of spatiality. Staying with the trouble, created by their opposition of forces of time and space, these experiments were both fascinating and vexing, in their status as science and as art.

How Moving Maps Work

Jacque Bertin's warning in 1967 about the "overwhelming" qualities of movement in cartography would prove too tempting for the neo-Robinsonians. Cartographers and geographers, engaged in a project to improve the computational tools by which digital maps could be produced, were also (in the wake of Robinson's dissertation) necessarily honing their methods of evaluation and innovation. Indeed, the double embrace of technique and testing would characterize mid- to late twentieth-century academic cartography. Howard Fisher of the Harvard LCGSA realized this as he attempted to enroll more geographers into his project in the late 1960s, worrying the psychological aspects of perception with digital maps would need further study. It seemed that innovations in the field were too numerous and too rapid. Fisher and the LCGSA would move on to experiments with color, new database structures, new methods of symbology, and the steady, deliberate work of testing would be built directly into the process of development. That maps had a particular look that would make them more functional and effective objects of communication would require more research into their design and operation.

Behavioral cartography became solidified as a field of inquiry.[48] In addition to a focus on the science of functional design of static maps, cartographic scholarship would excavate the qualities of movement and maps that move, by attempting to isolate its composing parts in order to further measure their specific effects on perception. Much of the contemporary elaborations of this approach take root at Penn State University with key scholars such as Cindy Brewer, Alan MacEachren, and David DiBiase (all advisory descendants of Robinson[49]). The computational opportunity for digital mapmaking would rapidly expand through the late 1970s and the 1980s, alongside the solidifying interests of new corporate players in the development of GIS software. New techniques and software for geographic representation would thereby necessitate new testing, to understand how to register their effectiveness and market their impact.[50] In 1995 MacEachren would write *How Maps Work,* an exhaustive treating of the map and how it operates. I suggest that the title of this text indicates an intensification of (and not a significant departure from) a continuity established within Robinson's *The Look of Maps.*[51] And while MacEachren maintains that this work attempts to dispel the strict and sterile implementation of the map communication model by introducing contextual factors in the operation of map reading, the motivation (toward the scientific study of effectivity in mapmaking) is largely shared.

What behavioral cartography would come to emphasize in the field was the heterogeneity of the practices of map reading and of map readers themselves. Certainly the map was still an important medium for development, but to take functional design seriously would be to actually alter the standards and conventions of mapmaking in alignment with the research of map-reading behavior. That the map reader would no longer be considered universal was perhaps the most significant contribution to the legacy of Robinsonian thought. Research on map design, according to Dan Montello, would create a "detailed analysis and vocabulary for describing the varied tasks of map users and producers" and further the development of new map innovations and interactions "in terms of its effectiveness for helping people understand the world."[52]

Movement, as maps that move, would become an obvious area of map design research and map development. Undoubtedly, animated cartography produces a powerful medium for helping to understand the complexity of the spinning world. After all, the world moves—why not our maps? However, just as the addition of the possibility for using color in digital mapmaking vexed Howard Fisher and other map design researchers of the late 1960s and 1970s, the technical capacity for representing movement on the map (and having the map actually move and change) would present new registers of experimentation for proponents of functional design. As Mark Harrower would later reflect, "When it comes to designing animated maps, the bottleneck is no longer the hardware, the software, or the data—it is the limited visual and cognitive processing capabilities of the map reader."[53]

To get at the capabilities of the reader, MacEachren would suggest that movement was not just one additional variable in cartography, contra Bertin. Rather movement as expressed in cartography was composed of several visual variables, drawing on work with DiBiase, among others. He writes:

> What is probably true, however, is that on a dynamic map things that change attract more attention than things that do not and things that move probably attract more attention than things that change in place.[54]

While a subtle statement that may seem obvious for the lay map enthusiast, MacEachren's admission, more a *first law of maps that move,* indicates a number of implications. The study of map perception could no longer presume the viewing of maps. Much of map design research had presumed that the map reader would actually engage in a reading of the map! Instead, following the techniques of advertising and marketing, map reading would require techniques of gathering attention.[55] And the *quantity* of that attention, for MacEachren, would hinge on whether or not objects on the map that move across the surface would attract more attention than objects that change in place.

That aspects of the perception of movement might be subjective and generally beyond the attentive gaze of a map reader would be resisted by behavioral cartographers. "The behavior is 'the bottom line,'" as George McCleary Jr. would state.[56] The ineffability of movement in cartography would simply require more variables. MacEachren and his graduate students distilled six variables of animated maps:[57]

1. Display date: "The time at which some display change is initiated."
2. Duration: "The length of time between two identifiable states." Here, duration may include "scenes," or frames or states of the map where no change occurs.
3. Order: "The sequence of frames or scenes."
4. Rate of change: "The difference in magnitude or change/unit time for each sequence of frames or scenes (or *m/d*)." The movement of objects upon the map or even the map itself may change in rate, appearing to speed up or slow to a halt.
5. Frequency: "The number of identifiable states per unit time—temporal texture."
6. Synchronization: "The temporal correspondence of two or more time series."

For MacEachren and his map animation research team at Penn State, each of these variables could be characterized as numeric or non-numeric. In other words, while the moment in which a display change occurs and whether it was synchronized with other objects was considered largely nominal, the duration and rate of the change in the graphic could be measured at a ratio level. These variables of movement and their ability to be quantitatively measured would enable a new generation of cartographic study into the cognitive processes of map reading, an understanding that would tame the animated map and render it a more effective representation. New eye-tracking techniques would lend scientific rigor to these studies of map use and generate new applications of these techniques for industry. Still, much of this work was about the ability to understand the world, at scale and in situ. As Harrower and Sara Fabrikant argued in 2008:

> Better understanding of the human cognitive processes involved in . . . highly interactive graphic displays is fundamental for facilitating sense-making . . . Better understanding will lead to greater efficiency in the complex decision-making required to solve pressing environmental problems and societal needs.[58]

Scholarship around animated maps would investigate the potential for interactive maps in decision making.[59] In a project on "change blindness" published in 2011, Carolyn Fish, Kirk Goldsberry, and Sarah Battersby discuss the height of these methods, extending several decades of scholarship focused on functional design. In their map design test, users would view a dynamic choropleth map in graduated grayscale. After a time, an object on the choropleth map would change in symbology, and the users would be prompted to test their knowledge of which part of the map changed and indicate their level of confidence in that knowledge. Their research argues that a map reader's inability to confidently recognize change on a dynamic map would "undermine the effectiveness of cartographic animation."[60]

The question of how maps that move work, largely unchanged since the time of Robinson and Tobler, remains: How do we ensure that map readers and users receive the message encoded in the representation? What visual variables of movement can be resolved to better enable these new map users? Interestingly, the *subject* of such animated maps recedes into the background. Emergency response, vacation planning, personal navigation—the subject of such map interaction becomes largely unimportant in the wake of the measurement of efficient map communication. Maps become merely surfaces for use in decision making. Environmental contamination, social struggle, election results—these matter less in the fine-tuning of the map instrument. The trouble of the ineffable, of occupying neither science nor art, is the problem to be resolved. However, what might it mean to stay with the trouble, to maintain the "mysterious interaction between computer display and human visual processing systems," as Danny Dorling and Stan Openshaw argued in 1992 in the midst of the GIS Wars?[61]

Strange Concepts

MacEachren argued that the reader of a map deserved more study than was motivated by the map communication model. How maps worked depended on a number of variables and conditions. This includes the human capacity for attention:

> Cartographically, a key aspect of the way attention works is that initial views, if they take in large segments of a map, will be able to process only gross features. This processing will then guide the narrowing of attention to particular features and objects in order to examine details. Particularly in a visualization context, therefore, graphic design impacts upon the initial wide-scope global view of the map and may dictate what specific details are seen.[62]

Remarkably, the way in which human attention is seen to work parallels a notion of map scale. Our initial viewing of the map—that which captures our attention—requires a series of design decisions about what details on the map are generalized. After this initial viewing, MacEachren notes, the viewer may choose to examine further details expressed by the representation, thereby "narrowing" attention. The interplay between what might be termed *small-scale attention* and *large-scale attention* is produced by the map designer, with deliberate emphasis on the map reader. The widening and narrowing of attention to the map would be rendered systematic and thereby malleable.

That the working of the map is regarded as "processing" is important to note. Map reading and map use required certainty about known and unknown aspects of the map operation. This is symptomatic of a tendency within behavioral cartography to identify, isolate, and evaluate the variability of map reading. The rapid innovations in digital mapmaking technologies only tighten the bind between testing and technique. However, there are moments in the history of animated maps that highlight ruptures in this tendency.

I conclude this chapter by ruminating on two digital mapmaking projects of Geoff Dutton, a student and researcher at the Harvard Graduate School of Design in the late 1960s and 1970s. Dutton was a student of Bill Warntz and Mike Woldenberg, inheriting their tradition and curiosity of social physics and spatial analysis. In a class project from 1969

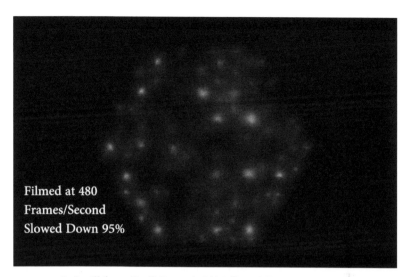

FIGURE 18. A still from Geoff Dutton's 2015 film of *Hexagonia*, a mapmaking project from his course with Mike Woldenberg at Harvard in 1969. Courtesy of Geoff Dutton.

titled *Hexagonia*, Dutton created a piece of electronic cartographic art, testing out central place theory in a nest of three orders of lights, representing interacting cities (see Figure 18). The work attempts to adjust our vision to the regularity of human life, order amid chaos. While the film of this work was created in 2015, the original hardware explored the beauty of the science of place. Its installation as part of the fiftieth anniversary of the LCGSA allowed viewers to pause and consider the ways in which the artistry of spatial representation was co-constituted with spatial science.

Dutton draws this work forward in *American Graph Fleeting* in 1979, the first animated map hologram. Figure 19 shows a promotional picture of the hardware, which displays a short map animation. Created using software produced by the lab at Harvard, the animation represents a beautiful, rotating three-dimensional drawing of population density in the United States. More than an artifact for precisely measuring the effectiveness of cartographic design, these animations illustrate what Rancière identifies in the troublemaking of cinema: "The art of cinema cannot only be the deployment of the specific powers of its machine. It exists through the play of gaps and improprieties."[63] Danny Dorling puts this differently, although just as aptly. "Cartographic animation is a strange concept. To animate means to create the illusion of movement."[64]

FIGURE 19. Dutton's *American Graph Fleeting,* produced at the Harvard LCGSA in 1979, was perhaps the world's first (and only) animated map hologram, viewable in the round. Photograph taken in 2011; courtesy of Geoff Dutton.

These two projects by Dutton demonstrate the strange concepts at the heart of animated maps. As experimentations, *Hexagonia* and *American Graph Fleeting* put a fine point on the shortfalls of the received cartographic tradition, that the impact of a map that moves resists the objective principles of functional design. In other words, as Harrower and Fabrikant argue, perhaps "there is the very real risk that mapping technology is outpacing cartographic theory."[65] Other theorization is necessary. I propose that the scholarly gaze of cartography might do well to consider the affectual dimension of maps that move, with the rich thought found within discussions of cinema.

The mystery of map animation may lie in its unresolvable tension, neither art nor science, of fixing space and opposing time, of equidistant instants and the movement-image of film. Perhaps the potential of maps that move is precisely in their uncanny appropriateness for our space-times. And this is where Massey might help direct such a critical cartography of animated maps. In the section of *For Space* titled "Falling through the Map," she offers her own interpretation of how maps work: they give "the impression that space is a surface . . . the sphere of a completed horizontality."[66]

Instead, to abandon the opposition of space to time allows us to think about maps of spaces that are "always unfinished and open." However, this is the problem; she continues, "Loose ends and ongoing stories are real challenges to cartography."[67] The static map tends toward closure, space as stasis. Perhaps maps that move might mobilize design to think the intervention of cartography differently, as shifting the ways the world is experienced and represented. Animated maps might be such interventions for liveliness, if we allow their surprise and disruption.

I suggest an evolving question at the center of map design research is where to place aesthetics. In the realm of a series of conversations about cartographic efficacy? Rather, what if this scholarship and animated map tinkering mobilized design as the way in which to think the intervention of cartography into the ways the world is experienced and represented? Academic, behavioral cartography knows, and has known for decades, that it should look beyond its immediate fields to better understand the reading of maps that move, that "much of the current literature on animated maps has been written without regard to parallel work in animation from beyond our disciplinary borders."[68]

To understand movement as the mapping of the trace of the map that moves is to attend to the powers of attentional capture—that most peculiar psychopower. The stakes are high. Massey calls on de Certeau, in his discussion of stories,[69] as time was transformed "into a quantifiable and regulatable space," "to the normality of an observable and readable system" where "surprises are averted." Massey closes this section, "We do not feel the disruptions of space, the coming upon difference. On the road map you won't drive off the edge of your known world. In space as I want to imagine it, you just might."[70] Maps that move challenge these transformations catalyzed by representations of spatial relations. These animations conjure the presuppositions of the map graphic and its perception. The intensity of the vibrations created in their wake are only ever recognized in a state of movement. We attend to them variably and sometimes not at all.

CHAPTER FOUR

Attention

Memory Support and the Care of Community

> It is far too simplistic to say that hyper attention represents a cognitive deficit or a decline in cognitive ability among young people. . . . On the contrary, hyper attention can be seen as a positive adaptation that makes young people better suited to live in the information-intensive environments that are becoming ever more pervasive.
>
> —N. KATHERINE HAYLES, *HOW WE THINK*

> And this is occurring because the solicitation of attention has become the fundamental function of the economic system as a whole, meaning that biopower has become a psychopower.
>
> —BERNARD STIEGLER, *TAKING CARE OF YOUTH AND THE GENERATIONS*

If how maps work has much to do with the attention of the map user and reader, as cartographers of the late twentieth century were tempted to find concrete measure for, then the shifts in the conditions for attention must increase in importance for those committed to new lines. To intervene, to draw a line, often requires an audience—those yet to be engaged, compelled, entertained, informed. Indeed, several decades of cartographic scholarship drew on the techniques for the crafting of attention that were being developed outside of the field. For Arthur Robinson in the 1950s and 1960s, that meant print advertising. For cartographers of the 1980s and 1990s, that meant documentary films, cartoons, and computer games. For contemporary mapmakers,[1] that means mobile app design and responsive and interactive web browsing experiences. When maps work well, they work because to what we attend in using them was thoughtfully and explicitly designed.

However, as chapter 3 suggested, there are intangibles, the ineffable, and multitudes of presuppositions that compose the workings and failings of the map. The rapid proliferation of online digital media further complicates the functional design mantra of mainstream cartography. The human capacity to pay attention is the explicit target of these changes in media, and both popular and academic presses increasingly point to the potentially negative ramifications for lives lived online. Hayles and Stiegler, in the previous epigraphs, mark a particular tension around how to respond to the incursion of digital media into everyday life. Hayles adopts a more productive reading of hyperattention, while recognizing the weight of retentional systems (such as long- and short-term memory) that are being reworked. Hyperattention effectively prepares individuals for the realities of living in current digital culture, she argues. For Stiegler, however, the rapid reconfiguring of retention demands a more urgent (and more foundational) response.[2] An industrial model fashioned around attention capture and control has managed to coordinate a range of media, from radio and television to the Internet, in all its guises as Web 1.0 or 2.0, collaborative and distributed, the cloud, among others. What is needed then, Stiegler continues, is nothing short of a new economic system.[3]

Regardless of these distinct perspectives on shifting retentional demands, the pervasiveness of technologies that target attention would indicate that as geospatial technologies align with trends in consumer electronics, scholars of GIScience must begin to understand their work as part of a general mediatization of everyday life, which includes social networking, location-based services, and microblogging web-based tools.[4] In other words, the work of GIS and mapping is media work. In the discipline of GIScience, this statement is well illustrated by the tradition of community-engaged GIS and participatory mapping. Increasingly, to support community-based organizations (CBOs) with GIS and mapping is akin to offering media support.[5] From Bill Bunge to Sarah Elwood, the work to partner with communities, to map their neighborhoods, is about developing specific capacities, peculiar attunements, to alter to what we attend and how we pay attention.[6]

After overviewing my own efforts to promote community-based mapping in the undergraduate classroom, I analyze the broader implications for such media support with particular emphasis on attention. Stiegler suggests that technologies that target attention can be thought as systems of care. Taking care then, as in to attend or to pay attention, requires an elaboration on a Foucauldian notion of biopower toward that of psychopower: "that contemporary power technologies no longer mainly aim at disciplining bodies or regulating life-processes, but at controlling and modulating consciousness."[7] Stiegler insists that Foucault's later work on governmentality, as a cultivation of biopower through social regulations, could not take up the rapid proliferation of marketing and advertising technologies. His argument places telecommunication systems within a continuum of technologies that enable the necessary retentions that form: (1) an individual's perceptions and observations, as primary retentions, and (2) an individual's memory, as secondary retentions, as well as (3) collective memories, passed down through generations, as tertiary retentions. These three retentions, Stiegler argues, extend the constitutive processes of human culture itself.[8]

Advertising technologies, and the cultural industry more generally, work to short-circuit these retentions. These disruptions can be productive; however, Stiegler tends to emphasize the destructive and toxic nature of these moments of rewiring.[9] As Sam Kinsley suggests, the novelty of the industrial retention of everyday life is located in how these systems begin to predetermine social relations:

> The emerging industrial apparatuses for the capture, storage and transmission of memory demonstrate a potentially sizable shift in the ways in which we govern, negotiate and understand our collective life. On the one hand, they place the potential for an extraordinary level of control over what is remembered, how it is remembered and what influence this can have on contemporary socio-spatial experience—where we can go, with whom we communicate and so on. On the other hand, these "mnemotechnologies" are beginning to open out not only the access to but also the creation of shared, collective knowledge across a diverse spectrum of lives.[10]

The apparent urgency then is not to resist the technologization of memory (indeed this is intrinsic to human life), but to recognize the centrality of techniques of retention, and the rapid ways in which these techniques may become toxic—when these techniques predetermine collective life.

CBOs are but minor players in these retentional strategies, but their practices highlight the effects of the attention economy and the constrained potential for transformations through university–community partnerships with mapping. Stiegler's critique of the long-range effects of attention control technologies, what he terms "psychotechnological systems of psychopower," where human culture hangs in the balance, is a useful perspective to consider the conditions through which new lines might intervene.[11] One does not need to adopt the entirety of Stiegler's argument on the derailing of neural capacities for the production of culture to recognize the intense demands on thought and action wrought by digital media.[12] Community-based mapping can and must respond to these challenges. This begins by recognizing the ways in which mapping shares in these retentional systems.

Media Support Systems

Recognizing GIS support as media support brings new opportunities and challenges for community-based engagement. This recognition begins from two presuppositions. First, the notion of digital spatial technologies is expanded beyond the rather narrow set of systems, actors, and industries conventionally signaled by the use of the acronym *GIS*: these specific systems and the maps produced are just one part of a growing set of new spatial and social media. Second, the efforts to engage in collaborative and critical mapping practices, the hallmark of the critical GIS agenda, must be situated within our current moment, where digital information technologies are seemingly ubiquitous and increasingly pervasive. Leszczynski is instructive here; the point of this recognition is not to assert a singular genealogy of geographic information systems as emergent from wider developments in media. Rather:

> Epistemologically claiming spatial media as "media" directly asserts their materiality, and in so doing serves as a basis from which to grapple with the socio-spatial effects and significance of these technological phenomena through opening up the possibilities for engaging them in terms of ontological conditions of mediation.[13]

In other words, to know and think GIS support as media support, then, is to recognize these developments (with different techno/social histories) *as mediation*. The question of how to support CBOs with mapping becomes one of the conditions for spatial mediation, and more precisely, of how to constitute modes of attention.

Societal challenges resulting from the proliferation of code and computing condition the potential of GIScience to realize its scholarly as well as academic agendas. While the scholarly agenda of GIScience is to understand the changing capacities for geospatial data in the midst of pervasive locative media, its academic agenda is more Machiavellian—to continue to assert the centrality of geospatial technologies in understanding and coordinating everyday life.[14] However, the GISciences are not, nor were they ever, hermetically sealed. They recognize their role and remit, if not always, their responsibility. By placing innovations in mapping within the shifting conditions of digital information technologies, I argue that practitioners, scholars, and students of GIS are better equipped to adapt, respond, and take responsibility amid the complex relationships constituted by the interplay of technology and culture.

Therefore, a renewed commitment to critical GIS as a vehicle for community engagement is increasingly confronted with the following question: What are the social implications of persistent change in digital media, or the way in which updates to software and shifts in online functionalities rapidly iterate? I take up this question in three moves: First, I reconsider the conditions and remit of a critical GIS research agenda amid the advance of a promiscuous digital culture. Here, I suggest that so-called critical approaches to GIS and mapping must not remain silent on general technological shifts and intensifications in society, as forms of spatial mediation. Second, I overview efforts to constitute university–community partnerships with digital spatial technologies, to illustrate how these classroom projects intersect qualitative research into the everyday data and technology practices of community engagement. Finally, I draw out what I see to be the pressing concerns and opportunities of this evolving agenda, particularly focusing on challenges of the attention-work of community-based organizations in the wake of persistent change in online digital spatial mediation.

Conditioning Critical GIS

If critical GIS is a tacking back and forth between technical practice and critical practice, then it is this kind of technopositionality that enables a different narration of persistent change in digital media.[15] Working with digital media in partnerships with community organizations enables a witnessing of the vulnerabilities associated with an evolving digital culture.[16] These vulnerabilities condition the application of critical mapping approaches. As these collaborations with GIS unfold, the work of critical GIS becomes partially media strategy and support, to leverage the legitimacy afforded mapping to promote awareness and engaged thought in alignment with the CBO. If the point of these partnerships and collaborations are to help CBOs gain attention, and ideally legitimacy in their concerns, then an understanding of the milieu into which such interventions are staged is significant.

Practicing critical GIS with community partners is to engage in media production, and enables an active process by which issues of "representation" and "radical intervention" are not stuffy concepts reserved for abstract discussion, but are the flashpoints for theory-action.[17] Criticality in mapping embraces these moments of contact between the map, the territory, and the map reader, but not by considering these as separable entities of calculated interaction.[18] Maps are representations, and as such, they intervene; they leave a mark. A critical mapping approach revels in the space between representation and intervention. It brings methods to examine and understand these moments, and does not seek to minimize their implications or cloak their position. Committed to anti-opacity (which is different from a fictive transparency), the agenda of critical GIS enables a discussion of the mediating effects of mapping and spatial media while remaining attentive to the more technical decisions of map production and spatial analysis.

Mapping technologies have shifted alongside broader changes in digital information technologies. Just as the ubiquity of digital information technologies has dramatically altered everyday life for some, so has the proliferation of social media augmented with geographic technologies in creating further splinters in spaces of public engagement.[19] Indeed, since 2005, with the explosion of user-generated map mash-ups with the Google Maps API, social networking tools have also undergone significant revision—with the move toward mobile as the primary development environment and the intensification of location-based services.[20] It is impossible to ignore these developments as somehow entirely distinct from the developments that underlie the GISciences. Increasingly spatial media are interpellated by social media. Their logics of development, marketing, and use overlap and align.

Consider a few recent examples of the technocultural transformation of everyday life. The Internet meme has been heralded as a leading form of viral engagement,[21] a mash-up of various cultural and commercial content, while corporations compete for control of our attention, as seen in a recent marketing campaign by Microsoft to unseat Google as the primary Internet search engine. New human-computer interactions further lock in our ways of being in the world,[22] as software like Apple's Siri constitute spaces where the friction between the material and the seemingly immaterial are made insignificant. These interactions are no longer exceptional and instead signal a public increasingly demanding social-spatial mediation of everyday life.[23] And these devices are data hungry, leading to new opportunities for innovation in digital infrastructure, as was seen at the 2012 SXSW gathering, where homeless bodies were enrolled as wireless infrastructure.[24]

Therefore, I suggest that engagement in critical mapping, to draw new lines with collectives, and in the study of mapping as a cultural practice, demands an attention to the myriad intersections of capital and innovation, devices and desires, imaginations and urban governance. This attentiveness can begin, I argue, by recognizing systems for geographic information as but one part of an expanding digital culture.[25] Engagement with geographic information technologies is not, and was never, wholly separate from the industrialization of collective memory and the attention economies that produce and organize to what we attend. However, the scholarly agenda of GIScience has been distracted from these broader industrial transformations.

Technological Engagements

Conventional GIScience is being confronted with new technologies and new forms of data.[26] This is not necessarily forming a radical break, but is placing new societal importance on geographic representations and constituting new pedagogical challenges for GIScience training. These engagements are inextricably bound up in the *practice* of critical GIS. As such, I draw forward these epistemological and ontological critiques of GIS as an object, in order to imagine GIS differently. To take geospatial technologies as an object of study enables inquiry into the social and political implications of this specific software and hardware, as well as the habits of thought promoted by the use of such tools.

For instance, there are distinct challenges around privacy that become particularly pressing once GIS is recognized more broadly as media. Elwood and Leszczynski analyze the various discursive strategies employed around the question of privacy and the geospatial web, while Nancy Obermeyer targets an assumed voluntarism at the heart of the geoweb, noting the growing ways in which geographic information is collected through the use of consumer electronics and social media websites.[27] Indeed, "the geoweb forces us to think beyond a singular technology (GIS) and its primary representational output, the map."[28] These technologies are necessarily more, exceeding our conventional understandings of the relationship between maps, territory, and reader, while introducing new problematics as these technologies expand into many conspicuous and inconspicuous facets of everyday life.

Further, a political economy of the geoweb views continuities between the map that serves the interests of the state and emerging geospatial data.[29] Leszczynski argues that a "complementarity" exists between the mapping regimes of the state and the market, complicating any simple delineation between the state and the market, between "roll-back" and "roll-out" narrations of neoliberalism.[30] Positioned within critical GIS, this scholarship works to situate the technologies signaled by use of the acronym *GIS* within much broader political and economic conditions—including the uneven neoliberal restructuring of governance and the rapid commercialization of digital spatial information.

An expanded critical GIS research agenda places this work more directly in conversation with broader research within critical technology studies, conducted both by geographers and more generally by media studies scholars. That the constitutive relationship between technology and society is spaced is a long-held tenant within geography, and most recently elaborated by geographers attuned to the digital as an intensification of this relationship. Geographers temper the notion that technology determines space and spatial relations, by recognizing the multiple facets of sociotechnical relationality.[31] To recognize geospatial technologies as productive of these relations effectively widens the discussions within GIScience to include not only the social and political implications of GIS (a hallmark of the GIS & Society agenda), but also the proliferation of digital media more generally.

While an emerging literature within the digital humanities and critical media studies examines demands of attention by current digital technologies,[32] discussion of geospatial technologies (such as location-based services) as conditioned by these demands has largely been absent. The implication of this absence has meant that participatory engagement with these techniques and technologies have underexamined the noetic, or the conditioning of thought and focus that comes with digital technologies. The paying of attention to attention itself is an important consideration in technological engagement with geospatial technologies, and this is evident in current work with community organizations around the use of digital spatial media, as I examine below.

CBOs may be quickly overwhelmed in a rapidly unfolding attention economy, where Stiegler argues cultural industries dominate, including TV and radio conglomerates, the entertainment sector, and digital social media, albeit with different speeds and histories of involvement. Such cultural industries manage this through attention control, according to Stiegler, by directly targeting the human capacity to pay attention. Federica Frabetti writes:

> For Stiegler every technics (for instance, pottery) carries the memory of a past experience; but only mnemotechnics (for instance, writing) are conceived with the primary *purpose* of carrying the memory of a past experience. In Stiegler's argument, the emphasis is on the aim, or end, of different technologies: some technologies are conceived just for recording, others are not.[33]

Indeed, as Frabetti continues, software is not just about recording but about making "things happen in the world."[34]

Stiegler encourages an investigation of these attentional demands across a range of media. For example, James Ash has examined how first-person shooter video games produce and alter notions of temporality.[35] The technicity of such video games enables multiple understandings of the passing of time, and highlights, for Ash, the ways in which technological objects are constitutive of being and becoming. These objects and the techniques that shape them have technicity—or the capacity to constitute beings. This technicity is also fundamentally material, in the ways in which the shaping of attention is predicated on microelectronic devices, an argument furthered by Sy Taffel.[36] In other words, these devices also have ecological costs. Further, attention control can be examined as a process of transindividuation; Ben Roberts examines free software as a process of public making, through which individuals are constituted as people who tinker and invent.[37]

This approach, of interrogating the productivity of digital information technologies, situates scholarship on attention as an examination of the operative work of power. I suggest that technological support of CBOs is situated within these power dynamics, and therefore requires participatory work with GIS to consider the attentional demands of digital media. The ongoing work of partnership with GIS is increasingly confronted by the effects of an attention economy that Stiegler examines with great urgency.

Mapping to Learn

Classroom partnerships with CBOs around the use of GIS enables a range of student mapping projects, conducted with community–partner direction and oversight. The maps demonstrate a commitment to the process of mapping to learn, to enroll mapping not as the final, static product of geographic investigation, but as the mile marker of an unfolding partnership. These partnerships have culminated into what I have called "GIS Workshop," a capstone GIScience course designed for advanced GIS students.[38] GIS Workshop has similar models of classroom partnership as that facilitated by the Syracuse Community Geography program that facilitates community-led research projects, akin to public participation GIS.[39] While the mapping projects begin and end with the course, the partnerships are sustained through continued yearlong follow-ups with CBOs to establish strategies that make best use of student expertise and university technology resources. See Figure 20 for a selected list of these partnerships.

These projects demand an expansive set of technical facility while providing students with a cross-disciplinary approach to better understand a diversity of human–environmental conditions. Students map in order to learn about their local communities, and many use the classroom-based

Community Partners	Classroom-Based Support
Nonprofit Healthcare	Better understand healthcare service areas, analyze, and represent the expansion of their spatial footprint
Urban Community Development	Map predatory lending activities in the city, analyze the ways in which more unsavory capitalistic activities have targeted areas of poverty
Nonprofit Food Support	Track and map grocery stores, community gardens, and restaurant inventories, analyze and represent local food systems
Rural Community Development	Map the social implications of post office closures in rural Appalachian counties, document the myriad relationships between the post office and small communities and the adverse effects for an aging population
Tourism Development	Map opportunities for historical tours
Animal Services	Map the locations of animals picked up by municipal animal control and analyze the implications for reduced city services
Preservation Advocacy	Analyze and represent multidecade efforts of historical preservation efforts in a rust-belt city
Environmental Advocacy	Analyze and map volunteer-generated data about water quality in/near the Kentucky River

FIGURE 20. Examples of community-based critical GIS partnerships, 2010–13.

projects as a springboard for other modes of engagement during their college career and beyond, as some projects turn into volunteering opportunities and internships.

While engaging in these projects since 2010, the students and I have witnessed how data and the representation of data figure into the everyday practices of nonprofit organizations. The workshops are increasingly impacted by the persistent change of digital media, namely, as new spatial media creates alternative mapping practices. Therefore, the work of building and sustaining these partnerships increasingly demands a more general media strategy that begins through an inventory of the ways in which CBOs use digital information technologies. In what follows, I present some of my observations and interviews with community partners as they discuss the current challenges of working with digital media. More than thirty community-based projects, with more than twenty community partners and more than seventy-five graduate and undergraduate students, have been facilitated in GIS Workshops, in four semesters, 2010 through 2013.[40] Being attuned to community organizations' needs for spatial analysis and representation, as part of a community-based critical GIS agenda, means also recognizing the myriad ways in which these organizations use digital information technologies.

Digital Media Work

A long-standing community organizer, Tanya Torp, was featured in the *Lexington Herald-Leader* to highlight a broad effort to tackle issues of food insecurity in the northeast end of the city—the location of historically black and lower-income neighborhoods:

> The effort takes a network of volunteers to "glean" and distribute produce to neighborhoods where it is needed, farmers and farmers' markets that are willing to donate leftovers and neighborhood captains like Torp who help distribute the produce and build communities.[41]

Increasingly, Torp's everyday work in communities is the work of networking and distribution, of connecting individuals and organizations, of building community through these connections. Her work—like that of many of the individuals I meet and partner with through GIS Workshop—is increasingly dependent on mediation by digital information technologies.

At the outset, however, many partners recognize that while digital tools are useful for fundraising and creating awareness, the bulk of community outreach necessarily requires face-to-face interactions.

> And that's the thing in our neighborhood . . . being able to reach people. You've got to go old school and knock on doors because a lot of people don't have access to a computer.[42]

This community organizer, who focuses on self-esteem among young women in Lexington's East End, knows that outreach is primarily "old school." Indeed, during my meetings with community partners, there is a general sense that while digital information technologies are certainly being taken up for the management of volunteer resources and for communication with constituents, many of the recipients of the services provided by these organizations do not have basic access to digital information technologies.

Community partners suspect this divide. Nonetheless, their everyday engagements with the technologies speak to particular challenges in the rise of digital cultures. For some partners, social media websites such as Twitter and Facebook are useful to gather information and perhaps less useful in getting information to the individuals they service:

> I was wanting to start a Twitter account for us, but I think Twitter is more useful for us in bringing in information. Because farmers don't Twitter.[43]

For this staffer at a nonprofit that advocates on behalf of Kentucky farmers, the work of communicating with communities sometimes means posting information on doors. For other nonprofits, national or regional fundraising campaigns will often require their participation in social media, although those they serve are likely not using these platforms:

> When it comes to social media stuff, I personally don't like Twitter, so I only tweet when I have to. It's usually during the GoodGiving Guide challenge because you have to tweet and Facebook to be a part of it.[44]

The rise of digital culture is unevenly experienced, a well-documented phenomena within geographies of ICTs.[45]

Furthermore, the digital information technologies that are used by community partners are subject to technological shifts and evolutions. Many of these changes and adjustments are frustrating, if mundane, and yet demand vigilance by partners who want to remain connected, who want to make certain they have the latest and most widely used platforms for online presence—platforms that seem to be in a constant state of change:

> And then I ended up changing our Facebook page to a page for a nonprofit organization as opposed to a group. . . . So then I'm trying to move people and go: "This is gonna go away. Stop liking this!" . . . You have to just ditch it and start over from scratch, which is kind of what we did.[46]

Most commonly referenced by community organizations is the frustration of the forced shift from Facebook groups to Facebook pages, a slight change in functionality, as is demonstrated by this quote from an individual who works for a faith-based nonprofit that facilitates charitable food donations from local grocery stores. This shift in functionality caused this community organization, which relies on Facebook as a primary vehicle to communicate with volunteers, to "start over from scratch." And while this may seem like a minor implication for using "free" web-based resources, it underlines the potential consequences for organizations that utilize online social media corporations for their primary form of communication and engagement.

For organizations that must make budget-neutral decisions about their media strategies, it is difficult to know which digital information technologies are appropriate to adopt and which ones might likely "wither":

> So I don't know if there's a good gauge about which things to adopt, which things to pass while they wither on a vine somewhere.[47]

Website designs may eventually appear dated. Social networking applications may go out of business. Free functionalities may risk being rolled into paid subscription services. And when staffed by low-paid or unpaid volunteers, nonprofits may not feel particularly compelled to invest that time, when the knowledge required to maintain such online commitments may disappear with the frequent shifts in a primarily volunteer labor force.

For organizations that recognize the opportunity of personalizing their mission through digital information technologies, the topic of digital data storage is one of additional frustration. Many sites that store photos, videos, and documentation have free data storage under a certain threshold. As a nonprofit organization, the management of that threshold becomes part of the mundane practices that surround digital work:

> I've pulled the pictures out because of course we're using the free Dropbox. So, I've pulled the pictures out and I started putting them on Picasa.[48]

Here, a volunteer describes a decision to transfer digital content from one free service to another, as a way to manage the threshold for free online storage. CBOs accommodate the hassles of digital media, because access seems largely free:

> We have Facebook; we have e-mail. Every so often we send out stuff. Whenever there's a real need we send it out . . . I used to, every month, send out an update on e-mail, but our computers are so damn screwed up, it's hard. We go by what we get for free. And you know, as the Internet in the sky, that highway gets more and more jammed.[49]

In an environment of increased competition for a finite group of volunteers, CBOs necessarily engage in online social media as a way to market their specific mission. As a result, Facebook, Twitter, and even YouTube become the everyday tools to facilitate that communication and engagement.

However, CBOs must make decisions about where to best spend their energies:

> As a director, I do make time to be in front of a computer one or two hours each day. And that's a tension for me, because on one hand I understand that time in front of the computer can help time in the garden, but . . . If I'm not in a garden, SeedLeaf work isn't happening.[50]

For this director of SeedLeaf, the tension over this kind of digital media work is significant. Time spent on Facebook and Google Docs is in direct competition with time spent in the gardens. Indeed there are trade-offs between fully engaging digital information technologies, such as social media, and engaging in the core mission of the organization. If the director is not in the garden, "SeedLeaf work isn't happening."

For many of the CBOs in Lexington I have met with, volunteers, potential donors, and partner groups are kept informed through e-mail marketing companies such as MailChimp.[51] However, according to some CBOs, the free functionalities of these kinds of web-based marketing management companies shift, requiring organizations to either start paying for subscriptions to services or migrate their contacts and content to another marketing company with similar free features:

> You basically go with an email marketing company. And there's several out there. So you look through them all, but since we're small and have no money, we get the free ones. So then you spend all this time transferring your stuff from there, trying to figure out how to use that program.[52]

The frustrations of time spent learning new web-based programs to continue the infrastructural work of building followers, to connect and network, is a palpable concern. One partner discusses a program developed for nonprofits to help track food delivery:

> When we first started in 1998, [the program] only allowed up to one hundred clients free. Then you had to buy the software, fairly reasonably priced. It was still about $400. But I have found one, I just have yet to order it for one that is free for up to one thousand clients.[53]

When the costs of these subscriptions become too unaffordable, the work likely returns to a staff member, who is pulled away from more direct service:

> We have a company called Brewer Direct. They did do all of our e-mail marketing until recently, but they went up on their prices, so we were just like, okay, now I get to do it.[54]

While perhaps mundane, this work takes on particular importance, I suggest, in the context of increasing competition among systems of attention control.

The anxieties evoked by digital information technology work conditions the possibilities for a community-based critical GIS. The technological concerns of volunteers and staffers at community organizations must, therefore, be more fully considered as part of an unfolding process of technological engagement. Web-based software changes at a more rapid pace than desktop software on personal computers. For organizations, this means choosing among several tools, and engaging in that decision-making work continually as web-based tools shift their functionalities from free to subscription-based services. This work is compounded by personnel changes, as account passwords get lost, and the everyday work of maintaining digital information technologies must be relearned by new volunteers and new staff members. The tools that are used often have limited free functionality, causing some workers to be creative in the distribution of web-based content to manage things like free storage ceilings and the need for multiple strategies to communicate in an uneven digital culture. These concerns extend long-standing issues around equipment needs and bandwidth requirements for the latest iterations of digital information technologies, in the context of expanding impact and measurement regimes necessary to keep nonprofit CBOs funded.

Pay Attention

Persistent change in online digital media has specific implications for attention. This is illustrated by briefly examining the promoted features of MailChimp, the e-mail marketing company often referenced in discussions with community partners. These capabilities underscore the concerns and opportunities for the capturing of attention. MailChimp helps users to build a list, with custom forms and Facebook integration, to create a template, with web-based images and file hosting, and to send a specific campaign to that list, with segmentation by location, activity, and interest, with autoresponder bots, enhanced through social networking. Importantly, MailChimp allows users to track the results of their campaign, with automated reports, Twitter trending data, and Google Analytics integration. MailChimp enables a systematic and even automatic communication and then provides the calculative tools to assess communication campaigns—in order to best craft and control attention.

However, MailChimp is just one of the many digital information technologies enrolled by community-based partners to support the core work of their organizations. There are also sites used to gather information and connect with online social networks, such as Facebook, Twitter, and even Pinterest, as well as sites used to quickly publish information to free-hosted websites, like WordPress and Weebly. CBOs use digital information technologies for web-based storage of documentation, videos, and images, through sites like Dropbox, Google Docs, Flickr, and YouTube, as well as enrolling more specialized sites to manage financial and volunteer resources.

These sites necessitate an assemblage of account names, passwords, security questions, and mobile phone backups. They generate scores of automated e-mails to account holders, reminding them of new content in the network, new activity, new followers, new requests, and account privacy changes. In addition to these daily reminders, changes in functionality and changes in personnel burden CBOs with the mundane work of maintaining sites, and thereby maintaining visibility within diverse networks.

CBOs' struggles over digital information technologies are real challenges: to have and maintain a Facebook page, to tweet and to follow other Twitter users, to engage in the personalization of their campaigns using e-mail marketing websites. This is a struggle over capacity. And yet, as I am attempting to argue here, these everyday practices are more than a struggle for the capacity (both technological and personnel) to engage diverse publics. These struggles are more, both interior to these more technical concerns and external, situational, conditional. There are broader challenges that are productive of the conditions that give rise to the more mundane struggles of these organizations. In other words, these struggles over the maintenance of digital information technologies are but symptoms of an attention economy increasingly dominated by the cultural industry. The strategies to respond are not well traced.[55]

Attention Overload

Recognizing the pervasiveness of attention control highlights the challenges for strategic response. The persistence of change in online digital media has meant that short-range solutions offer only short-term resolution. Here, I would place efforts like the retooling of personnel in the use of digital media at CBOs as more immediately necessary but actually sustaining of the more negative aspects of the attention economy. Research in public participation GIS has long understood this tension in organizations with high turnover (particularly among more technically trained staff).[56] Other strategies leverage a range of immediacies and effectiveness, to include sustained access to digital information technologies, a stabilization or centralization of frequently used web-based functions, better or more university–community partnerships, or the development of new and tailored digital information tools for the nonprofit sector (or for the food-security organizations within the nonprofit sector or for the organic advocates within the food-security organizations within the nonprofit sector, and so forth). I could go on.

These strategies, while noble and incredibly useful within precise spacetimes, might instead be considered Band-Aids on a much more widespread problem of attention overload. Community-based work increasingly necessitates digital work. This work not only can pull the attention of organization staff away from their core mission, but much of this digital work is about fostering audiences through attention control. The work of building Facebook pages, following Twitter users, posting blog entries, and managing web-based content is largely the work of drawing people into the mission of an organization, to personalize the organization while promoting the agenda of the CBO. Persistent changes in online digital information technologies necessitate an organization's vigilance in the maintenance of their online presence. They must participate in this attention economy, and yet their participation may actually create new vulnerabilities. And as Massumi argues, "Nonconsciousness becomes the key economic actor."[57]

Community partners understand these practices as the necessary work of building and strengthening organizational networks and providing opportunities for further engagement and promotion of an organization's agenda. However, reconceptualizing this work as part of an attention economy overrun by the cultural industry perhaps underlines the complex interactions and potential implications of such attention strategies. Following Stiegler, this struggle over attention unravels the core of humanities' practices of exteriorization—of retention, collective memory, and the production of culture. One community partner underlines these new commitments, when asked how they stay in touch with constituents:

> We have a Facebook page, and we push a lot of stuff through that. We have a Twitter account, which we haven't used recently but we did use it a while ago. We also have the IdeaPost [blog] . . . We have a YouTube channel. We put some of the videos up of Ben de Jesus telling his story. We have a newsletter, a MailChimp style e-mail newsletter.

Digital information technologies, in extension of Stiegler's critique of telecommunication systems, rewire the conduits through which collectives are made possible, and further, how culture is produced.

Further examination of the use of digital information technologies by CBOs can yield a more complex understanding of the short- and long-range challenges of their utilization, a tracing of these nomadic practices to mobilize a collective. As CBOs move from website to website in order to increase cost savings, they are also enrolling new attention controls to build a collective. These technologies are thus both a problem and a solution. And the tradition of critical mapping as well as the broader GIS & Society movement is able to recognize the possibility of the enabling and disabling effects of psychopower, the pharmakon as the cure as well as the poison. Here, Stiegler draws on Derrida's reading of Plato's *Phaedrus*:

The original possibility of the image is the supplement; which adds itself without adding anything to fill an emptiness which, within fullness, begs to be replaced. Writing as painting is thus at once the evil and the remedy . . . Plato already said that the art or technique (techné) of writing was a pharmakon (drug or tincture, salutary or maleficent).[58]

The supplement, described here, that adds nothing and yet needs to be replaced is the pharmakon at its most toxic. This supplementarity may short-circuit primary and secondary retention (as observation and memory), but its most dangerous expression is in the transmission of culture (as tertiary retention and intergenerational sedimentation). While it is important to document the numerous ways in which digital information technologies as utilized by CBOs work to channel and control attention (becoming toxic), that the *technical* system of psychopower is a necessary aspect of the reproduction of humanity is the difficulty, the trouble we stay with. Therefore, it becomes more important to develop new practices that build forms of awareness that constitute collective memories and cultures of action, within and not without that technical system.

Sharing as Solidarity

However, a practical response to the attention challenges of persistent change in digital media is not entirely clear—perhaps unsurprising given the pharmacological framing of the problem. As both poison and cure, the "problem" cannot simply be resolved through technological disengagement, a rejection of *techné*. To return to this chapter's epigraph by Hayles, hyperattentiveness is an immediate resource to a generation of digital cultural workers. Instead, perhaps technological engagement requires an awareness of the *conditions* of thought-action, to better frame interventions with technology by being aware of the tendencies toward attention craft and control. William Connolly writes:

> Thinking is not merely involved in knowing, explaining, representing, evaluating, and judging. . . . To think is to move something. And to modify a pattern of body/brain connections helps to draw a habit, a disposition to judgment, or a capacity of action into being.[59]

To consider what this might look like in the practice of mapping with community-based organizations, I suggest three jumping-off points to hopefully continue the conversation.

First, research as to the digital information technology work of community organizations has made me think more long term about partnerships.

This means that work with community partners never really begins and ends with the GIS course. Instead, partnering has different speeds and volumes, and occurs within the context of an attention economy. To prevent the short-circuiting of retentional systems, Stiegler has highlighted the long-wave processes of intergenerational sedimentation—the passing along of memories (and culture) through the generations by an investment in technics, as care of the self and others. To engage community partners in the drawing of new lines requires a recognition of the privilege of the scholar, that interventions occur at rhythms that may exceed the university academic calendar.

Second, taking these partnerships seriously has meant developing a full range of technological inventories and digital strategies for building audiences and, thereby, collective memories. In other words, it is not sufficient to partner only around GIS or new spatial media, broadly understood. Instead, partnerships in order to build collectives require mediation, to leverage pervasive digital information technologies. The point is to pay attention to attention as an object, to cultivate *attention as care* through technological engagements to confront what Stiegler considers a "systemic carelessness,"[60] or, more profane, where "I don't give a fuck" *(je-m'en-foutiste)* has become a persistent affect toward societal (human, environmental, cultural) challenges.[61]

Finally, this research into the implications for persistent change in online digital media has underlined attention work as a key aspect of action. Recognizing this aspect places the work of partnering as part of that culture of action, to act on strategies for building a collective through recognition of the multiple aspects of the struggle for attention. This collective assumes responsibility in a new economy of contribution; according to Stiegler, "Responsibility is *shared* through attention formation, and this sharing is the grounding condition for solidarity."[62] What is needed is not necessarily new technologies or new technical practices to alleviate the anxieties of digital culture, but perhaps an awakening as to the ways in which digital information technologies, including new spatial media, capture our attention, an awakening as to how to foster, how to design, new attentional practices in mapping partnerships.

To engage questions regarding the reading of the map is to simultaneously ask questions regarding the conditions of that reading. Already this brings the drawing and tracing of lines in our location-aware society to a series of precipices. The state of the map in the 1950s and 1960s led to these kinds of certitudes by Arthur Robinson:

It is known what kind of rectangle is best, what kinds of layout or lettering can be employed (or avoided) in order to suggest (or not to suggest) such intangibles as stability, power, movement, and so on. There are widespread preferences as to colors, shapes, and designs that must be utilized to make a more effective presentation. The number of cartographic principles based on research and analysis in optics and psychology is surprising.[63]

Rather, what is more surprising is the continuity of this thought in contemporary academic cartography.[64] Indeed, while a behavioral cartography continues to advance questions regarding the effectiveness of maps, a broader understanding of the relational and affective events that compose the map yields some problems.

Those "essentially subjective" elements that Robinson diagnosed in the 1950s are also those elements that cause us to lean forward in our seats, toward the maps before us. They captivate and motivate thought and action. To ignore the attention economy and the affective moments of map events is to only examine the map under laboratory conditions. That maps move and create attention will require new lines, with a different sense of the neo-Robinsonian tradition—one that begins from the heart of the science of capturing attention: "if the cartographer is interested in the visual effectiveness of his [sic] map (as he should be), color of print and background is at least as significant in cartography as it is in advertising."[65] At least as significant, indeed.

CHAPTER FIVE

Quantification

Counting on Location-Aware Futures

> To this end, we will introduce tools to explore and understand cities that are based on the dualism in interaction between flows and networks, which are different sides of the same coin.
>
> —MIKE BATTY, *THE NEW SCIENCE OF CITIES*

> It seemed to them [eighteenth-century contemporaries of Adam Smith] that, just as the free flow of blood nourished all the tissues of the body, so economic circulation nourished all the members of society.
>
> —RICHARD SENNETT, *FLESH AND STONE*

I return to a previously mentioned quandary: the lines we draw, in turn, draw us in. This affective moment of the map event can be alternatively thought as a question of habit—of consciousness itself and the action, inaction, practice, and thought that saturate these moments. To ask questions about habit is more fundamentally to ask questions of being, according to Félix Ravaisson. He writes in *Of Habit* in 1838: "The universal law, the fundamental character of a being, is the tendency to persist in its way of being."[1] To recognize and seek to understand "being," Ravaisson continues, already presumes space and time. Therefore, by this thinking, in the disruption of the persistence of being, space becomes temporalized. The lines of the map register such disruptions, causing us to move, to attend, to care. But as I have argued, movement and attention largely exceed conventional thought in cartography and GIS. To think movement and attention together, and their powers of habit, I consider a third moment—quantification—and its role as the definitive map in a location-aware society. Richard Sennett and Mike Batty, in their quotes in the epigraphs, document different ends of the project of civilization and the economic power that comes of the disruption of the persistence of being and the temporalization of space. For Sennett, movement about the city was thought to enact good economic health for a society, paralleling eighteenth-century knowledge about the flow of blood in the body.[2] The configuration of streets and pedestrian walkways was about circulation. The design of the interiors of buildings and movement from the building into the street into town centers reflected this knowledge of the body and its habits. For Batty, new techniques and technologies of the network in society are elaborations of the flows that have always characterized urban life.[3] His use of the word *coin* is an instructive happenstance: as each side of the same coin, flows and networks make cents. While the techniques of network have proliferated, little seems to have changed in our mappings of the city.

New devices and techniques have emerged to better quantify an individual's movement, stasis, and even sleep. A discourse about the "smart city" applies these principles of measurement and quantification to the analysis, representation, and management of the city. The rapid pace of everyday life alongside increased individual access to digital geographic technologies in advanced capitalist societies has meant, for some, a broadened capacity to acknowledge, represent, and measure the movements and the habits of society. Perhaps a new quantitative revolution is upon us, as Elvin Wyly notes,[4] and along with it an evolved political economy that puts a premium on the digital dossiers of the masses, in service of an advertising and marketing agenda.

Quantification is a perennial bogeyman for the critical social sciences and humanities. While the concerns of numeracy, reduction, and logical positivism are often heard and reworked (for instance, see the special issue *Should Women Count?*, edited by Doreen Mattingly and Karen Falconer-Al-Hindi[5]), the disentangling of method from epistemology belies the industrial pressure on the making countable. In other words, making countable makes cents. Central then, to emergent mapping practices, is the proliferation of spatial data and the increasingly resolved footprint of human and more-than-human existences that are captured, counted, and controlled (and commodified!).

Cities are frequently invoked in these calculations, rethought as organisms while human bodies are quantified as systems. As such, the interactive opportunities and limitations for engagement, representation, and resistance are evermore significant. To understand the discourses that challenge and reorganize contemporary urbanism under the sign of quantification, I explore the parallel lines drawn among the rising consumer-electronic sector around personal activity monitors and the rapid visioning and speculation around smart urbanism. To be vigilant of the implications and affordances of quantification is to carefully attend to the dualities of the drawn line. A totalizing rejection of the force of quantification is too hasty. In her discussion of radio-frequency identification (RFID), Louise Amoore highlights this pharmacological moment of the location-aware society:

> The capacity of RFID to make us locatable is actually acutely ambivalent: we feel its potential to watch and to incarcerate just as we simultaneously feel it fulfill some of our desires and pleasures.[6]

Bernard Stiegler perhaps suggests that this ambivalence, between incarcerating surveillance and voyeuristic pleasure, is productive. Only when the extremes of the dualities become toxic is another relation urgently necessary.

In what follows, I examine recent developments in quantification, by tracing their edges. I suggest three ambivalences, three dualities: interoperability and propriety, competition and habit, and fashion and surveillance. As I proposed in chapter 1, quantification has emerged as a kind of new Mercator for the location-aware society—new lines for orientation and navigation, here from there. We greatly anticipate the affordances and shrug the constitutive implications of its use. (Just this morning, I received an e-mail from a campus administrator offering a Fitbit to anyone on campus who signed up for their online fitness program.) Therefore, beyond the quantitative-qualitative debates that forced unnecessarily nasty positions in the 1990s (mimicking the Hartshorne–Schaefer debates around discipline in the 1950s), this chapter examines the implications for thought and action, as quantification becomes the undeniable map for the navigation of selfhood and city. What capacities and habits are reinforced and developed through the implementation of these technologies and techniques? As a continuum of technologies that serve to both open and close engagement, these systems renew the specter of quantification, and yet our response, our critique, requires not renewal, but rearticulation.

Quantified Self-City-Nation

I proceed through technocultural critique.[7] The role of such critique of quantification is to examine industrial actions to better understand the conditions, both discursive and material, that enable stratification of the everyday. I call these particular stratifications *quantified self-city-nation* as a general theory for the assembly of technoscientific solutions to sociotechnical problems. But which comes first, the problem or the solution? My position in this technocultural critique suggests that this 'which comes first' question is already the wrong place to begin.[8] The point of the quantified self-city-nation is to conceptualize how these solutions are always in search of particular problems, and how this myopia—an inability to think outside quantification, or that quantification even has an outside—introduces some curious technocultures.

Quantified self-city-nation signals a multiscalar system of attentional control, where the organization of *an* individual body, its movement and stasis, becomes the mimetic resource for the organization of bod*ies*, both human and more-than-human. In other words, quantification is not just about counting and calculation and systems of empiricism, but about leveraging a discourse of predeterminacy, preemption, and a tilting toward the future. Digital devices that map the movements of a self become metaphors and figures to upscale these techniques for neighborhoods, cities, regions, and nations. Additionally, these devices for quantification are increasingly about the capturing of attention, more than the modulation of behaviors. As these technologies achieve greater ubiquity, their ability to have both greater resolution and personalization widens. With this comes a new register of problems that can be solved—through greater measurements and adjustments of systems. These advancements bring new opportunities for the channeling of the human capacity to pay attention, habit as consciousness. And it is the way in which these techniques slip across scales that piques my interest. How, and to what, might a body attend? A city? A nation? What are the possibilities for intervening within or channeling and focusing that attention? So many drawn lines. How then to map their traces?

To more abstractly identify the shifting arenas of body, city, and state management and modulation, I read across two specific developments—personal activity monitors and smart city technologies. If, as was explored in chapter 4, the possibility for intervention with spatial media is conditioned by an attention economy, then those techniques and digital devices that map knowledge of the self, the city, and the nation also operate through psychotechnologies of attention control. In other words, to follow Stieglerian thought, those processes that quantify the self-city-nation alter conditions for retention. Focusing only on the data produced and the innovations in techniques is to miss the tectonic process that reconfigures relations across the planet. Trees, no forest.

To be clear, I am not attempting to demonstrate empirically the concrete relationship between personal activity monitors like the Apple Watch

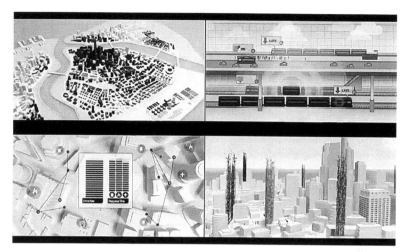

FIGURE 21. IBM's marketing campaign, *The Smarter City*, is announced in this video advertisement from 2010.

and the Fitbit and smart city software by companies like Esri and IBM. Undoubtedly these companies have their eye on broad attempts to expand their technocultural solutions to an ever widening social management and analytics machine. However, I am more interested in the use of suggestion to provoke and learn what might happen in the rub between these types of technological innovations and interventions, between the body and the city and nation. That these systems have yet to be fully realized—at their most idealistic and positivistic interoperability—is largely unimportant. Rather, my interest is in the discourse that permeates the quantified self-city-nation, where social and bodily regulation is made an obvious concern in the location-aware future and where speculation intersects with neoliberal regimes of political-economic life.

Consider two video advertisements shown in stills in Figure 21 and Figure 22.[9] IBM's *The Smarter City* marketing campaign has sought to rationalize a variety of urban functions, both public and private. In this one-minute video advertisement from 2010, a hypothetical city removed of texture and grit stretches out before the viewer, while a flurry of data networks and interfaces both monitor and direct movements and activities along the surface of the streets and within buildings. It begins, "The time to make our city smarter is now." The dynamisms of the city are multiple, awaiting only the appropriate sensors to both measure and represent their activities. It continues, "How do you demonstrate one decision affecting millions of people?" The urban fabric is availed to decision makers in this

FIGURE 22. Jawbone markets their personal activity monitor in this video advertisement, where a range of everyday activities are quantified and brought into comparison and competition.

way, and the heterogeneity of the urban experience is brought into coordination. In this vision, citizens are as much the agents of change as the sensors that compose the underlying infrastructure: "Become a citizen of the smarter city." Indeed, it is the way in which ideas about the smart city are composed that make this a rich site for critique. Smart cities, it would seem, are really about the behaviors of individuals, and the capacity for systems to alter those behaviors effectively.

In a similar dance of monitors and activities, Jawbone markets their personal activity devices to the mobile body. The UP by Jawbone is a device that measures not just activity, to include fitness and eating, but also inactivity: sitting and sleep patterns. The advertisement states, "See your sleep patterns. Track your goals. Wear it 24/7." Micro decision making becomes the primary driver of these devices, enabling a user to tinker with their bodily performance, with an explicit push toward greater movement and careful consumption, alongside thoughtful social interactivity. In 2014 Jawbone partnered with behavioral scientists to experiment with nudges and suggestions to see if these could actually change, significantly, the habits of users. Olson reports:

> Such intimacy underscores something fundamental going on. Consumers have been tracked, measured and prodded into action since the 1950s—psychology was, as any viewer of *Mad Men* recognizes, the very core of the modern advertising industry, with symbolism,

doublespeak and anxiety deployed for the first time as commercial weapons. Now the proliferation of connected devices—smartphones, wearables, thermostats, autos—combined with powerful and integrated software spells a golden age of behavioral science. Data will no longer reflect who we are—it will help determine it.[10]

The parallels between smart city software and personal activity monitors are pronounced. The double target of these techniques is the body itself and about what a body cares. Furthermore, as the above passage insists, these new developments are but changes along a continuity in the pharmacological interruptions of everyday life (from psychological studies in the advertising industry to behavioral studies in the consumer electronic industry).

How might one draw obvious comparisons between these advertisements? Movement and stasis, analysis and visualization, and decision making and behavioral change are the more obvious ways in which these proposals for technological innovation intervene in the everyday of the body and places inhabited together. Cutting across these advertisements is a kind of quantification, a valuing of numeracy where "being countable" is a precondition for the system. IBM's smarter city campaign is concerned with this kind of quantification as decision making. But, the success of such a "smarter city," the provenance of such an urban imaginary, demands involvement from citizens. One vision, which ad campaigns for personal activity monitors like Jawbone highlight, is to "quantify the self."

Making Autonomous

QuantifiedSelf.com, "self-knowledge through numbers," is a clearinghouse for how-tos, technology reviews, discussion forums, and meetups.[11] Self-tracking has become a beacon for a new kind of citizen engagement. Cities host meetups to develop and discuss these techniques. The meet-up page for Boston Quantified Self describes a "show and tell for people who are tracking data about their body."[12] Life logging, location tracking, behavior monitoring, and self-experimentation are the new technocultural practices of citizen engagement, a twist on neoliberalizing governance, where the individual becomes the key to social and urban transformation.

This form of self-care highlights a tension in the autonomous body and, according to Claire Rasmussen,[13] enables the possibility for a reconfiguration of power relations. To aspire toward self-improvement is to both be a docile body and practice resistance and reformulation through the malleability of the body. By making autonomous, these systems emphasize and glamorize personal responsibility and choice. Here, the scale of making autonomous shifts toward the nation. Indeed, the national census has long performed a quantification of the population, as a system of government. But the emergence in the United States of the Affordable Care Act brings to a point the work of making countable. Far from noncontentious, healthcare.gov catalyzed a specific moment in which the possibility for a more equitable system of health-care insurance access was refigured by the political right as an encroachment into everyday life and by the left as a woefully inadequate mechanism for social justice. What is most contentious *is not*, however, the making autonomous. Citizens are made more responsible and made to engage in more self-care. It is perhaps not a surprise that health-care insurance companies embrace the ubiquity of personal activity monitors—and the possibility of a more sousveillant health.

Representation is central to this form of making autonomous. Here, a kind of Cartesianism is implicit, wherein matter and perception are sources of trouble that must be kept sealed and separate. Maps are perennial culprits for this kind of trouble, as they cut apart social–spatial relations and become autonomous objects for the rationalization of space. As de Certeau writes, "What the map cuts up, the story cuts across."[14] For him, this process of automatization is visible in the play of figures across the map surface over the last few centuries, where:

> Far from being "illustrations," iconic glosses on the text, these figurations, like fragments of stories, mark on the map the historical operations from which it resulted. Thus the sailing ship painted on the sea indicates the maritime expedition that made it possible to represent the coastlines. . . . But the map gradually wins out over these figures; it colonizes space; it eliminates little by the little the pictural figurations of the practices that produce it. . . . The map thus collates on the same plane heterogeneous places, some *received* from a tradition and others *produced* by observation.[15]

The sea monsters, sinking ships, and celestial characters in the margins of medieval maps are identified as the quaint icons of a yesteryear cartography. The project of Enlightenment would remove the importance of these visual touches and mystic relations in cartography—and thereby, according to de Certeau, would make invisible the authorship, perspective, and position of the map.

This process of automatization has increased in speed over the last century. With the rise of urban data systems and the algorithmic governance made possible by these systems, it is perhaps easy to witness the slippages into urban science and smart city sloganeering that supports this scientific vision, as well as the moments of tangency with consumer electronics that conveniently peddles this vision. Observation and tradition are collated into representations that disguise their productions, and their ideologies. As Lefebvre writes in his provocations around such guises, "After all, who is going to take issue with the True?"[16] That maps lose those elements of their historical and material production furthers the map as a project of functionalism.

These suspicions are felt deeply in the discipline of geography, well before reactions to the quantitative revolution. That mapped representations were subjective, owing to their humanly creation, was the focus of J. K. Wright's oft-quoted article in 1942. As instruments of war, he writes, maps are autonomous objects for decision making as well as evocative objects for opinion and morale. And yet, he closes, "That you cannot navigate a ship without charts, however, does not mean that you can navigate it by charts alone. Rudders and helmsmen are also necessary."[17] This disclaimer, bordering on incantation, should bring us pause, in the rush toward biopolitical and psychopolitical software that seek to organize the everyday through quantification and visualization. Maps are both made as autonomous objects of representation and, through their proliferation as location-aware technologies, make objects and individuals autonomous elements in contemporary society.

Under the weight of these emergent technocultures, I ruminate on three conceptual pairs provoked by the rub between personal activity monitors and smart city systems. The first pair, interoperability and propriety, conjures questions about the implications for the production of standards under a quantified self-city-nation. The second pair, competition and habit, scrapes just below the surface of the measurement of movement to think about the productivity of these systems. In other words, what might these systems make possible? What do they *do*? The third pair, fashion and surveillance, interrogates the view and the making viewable in quantified self-city-nation, as an elaboration of a long-standing critique of representational technologies. These ruminations are not exhaustive but merely remainders in the rapid calculations of consumer electronics, managerial urbanisms, and the map. A renewed critical stance toward quantification should be positioned not as a rebuke of all numerical analysis and representation, as these kinds of strong abstractions are indeed necessary. Instead, our position on quantification should look to identify and challenge those moments of quantification that seek to alter the everyday through privatization and closure.

Interoperability and Propriety

In the wake created by her published research monograph on critical GIS in 1999, Nadine Schuurman would render visible a series of concerns that were presumed to be purely technical in GIScience. Interoperability was one such arena of debate—understood as the need to have systems that would allow the combining and sharing of spatial data, amid many platforms and formats for the creation of such data. That the basis of any solution to interoperability would require some type of standardization of meaning, would mean that pursuit of the technical means to combine spatial datasets is ultimately problematic. She writes, "Nailing down meaning is fraught."[18] Efforts to pursue interoperability on technical grounds are ill advised. Indeed, interoperability is an idealized solution to a technosocial problem.

Under the sign of the quantified self-city-nation, the concern for interoperability becomes urgent as the variety of sensors and systems of quantification of daily life proliferate. No nature, no body, no social structure seems able to resist this sign.[19] As a result, new industries, new consumers, and new form factors densify the digital mesh. Within the city, these changes prove too tempting for the practice of planning. In *The New Science of Cities,* Mike Batty traces a tradition of urban science and spatial analytics from the population settlement densities of Von Thunen in the nineteenth century to early twentieth-century social physics by George Zipf to the more recent law of online information sharing by Mark Zuckerberg, founder of Facebook.[20] Batty writes,

> New data begets new theory and already it is clear that the very focus of our interest is beginning to change. Most urban theory and indeed planning and design fifty years or more ago was predicated on radical and massive change to city form and structure . . . Planning was little concerned with smaller-scale development except its design . . . In short, the routine and short term were subsumed in the much longer term. New data and big data are changing all of this, as is the way the city is being wired and organised to deliver routine services.[21]

While midcentury urban planning and design fixated on the macro systems that characterize the life of the city—freeway interchanges, logistical operations, and the regionalization of the urban landscape—Batty suggests that more routine operations within the city are the current objects of planning and design. In other words, the availability of new networks and the flows these networks support actually change the rhythms of the city and the resolution of design and planning interventions.

While what exactly is new in this new science is perhaps up for debate,[22] Batty's advocacy of urban technologies is not. New technological infrastructure—largely digital and spatially aware—have enabled a recalibration of urban management in some cities toward the routine and everyday. Batty calls this the "short-term," enacting a systematic visioning of cities, "based on connections that manifest themselves as networks to deliver the flows of energy and sustain their moving parts."[23] The flows of the city, somewhat an inevitability for Batty, depends on the connections and networks to both create the pathways for their more efficient transmission as well as provide an agar that might allow urban mechanics to witness their activation. Metabolism of the city, thought this way, requires both connection and circulation, "different sides of the same coin."[24]

Anthony Townsend is also optimistic about innovation's combinatory effects that draw together the decaying physical infrastructures of cities with new architecture and new technological objects and industries. In *Smart Cities*, Townsend imparts a different lesson than Batty, however. He writes:

> We need to question the confidence of tech-industry giants, and organize the local innovation that's blossoming at the grassroots into a truly global movement. We need to push our civic leaders to think more about long-term survival and less about short-term gain, more about cooperation than competition. Most importantly, we need to take the wheel back from the engineers, and let people and communities decide where we should steer.[25]

For Townsend, the new capacity of urban dwellers to interact within the city—again, connection and circulation—begs not only the new science that Batty proposes (and all the shadow industries that support that science), but also what Townsend considers a "new civics"—an engaged and collectively minded citizenry that utilizes these techniques to remake the city, collaboratively.

The privatization and general opacity of these sociotechnical systems are sources of trouble, for Townsend. His address to these issues is to both involve the public at all levels and specifically condition citizens to tinker and develop these systems such that a thousand civic apps may bloom. In part, this is an educational mandate. However, Townsend's approach is also historical material, to "connect the schemes of the rich and powerful with the life of the street."[26] However, his vision is conditioned by a belief in the agency of individuals, a different myopia from Batty's claims that understanding the networks and flows of the city can resolve the sociotechnical problems of the urban. Townsend continues:

Until now, smart-city visions have been about controlling us. What we need is a new social code to bring meaning to and exert control over the technological code of urban operating systems. We need a new civics for the smart city that takes what we know about making good places as well as good technology, and shows us how to put them into practice. Only a sound set of guidelines will allow the designs for smart cities to emerge organically and to be shaped by the desires and choices of the people who must live in them.[27]

This "social code" would assert the power of the people to make change in their neighborhoods and feel responsible for and "in control of" the places in which they live. However, as decades of planning literature have debated, attempts at public involvement frequently ignore or reproduce the deep-seated structural inequalities that prevent such a civics.

Regardless, both technological and civic theses advocate a notion of interoperability—a fantastic interplay of human and nonhuman elements, an orchestration built from the local, indeed, with systems constructed specifically to understand the short term, the routine, such that a long-range vision can be sustained. Flows, networks, and especially engagement require a certain degree of coordination in this vision, and yet the implementation often falls short. What Townsend marks is an opportunity to leverage the pervasiveness of digital sensors and interactive devices to bring about new collectives and a new civic culture. Such an orchestration, however, would require an openness to interoperability—which too often is at odds with monetization. The idealization of interoperability requires propriety, as such a standard of behavior where sharing and civility supersedes the potential profit from any specific solution to a sociotechnical problem.

How then might we understand this *propriety* in the wake of *proprietary* systems and standards that work to close our quantified self-city-nation? That I cannot access or share the raw material of my personal activity monitor or that access to Esri software for urban analytics ranges in cost from $500 to $11,000 underlines the potential pitfalls of the quantified self-city-nation. To create the primary, ideally open, standard to measure, analyze, and represent multiscalar metabolisms, these corporations are fueling a new arms race over self-knowledge from the body to the nation. Data formats and metrics thus are far from inconsequential and become a neoliberalizing vehicle for the reterritorialization—as a dispossessing possession—of the body, the city, and the nation.[28]

Competition and Habit

The figuring of the city as an organism is an imagination resting on movement. As I discussed in chapter 3, a fixation on movement in geographic representation takes hold in the mid-twentieth century with animated cartography. However, what might be made of stasis, of a city staying in place? To think the city as an organism is to assume a city that expresses life through movement. Cities, as organisms, that are not in movement are in poor health or dead, the analogy goes. Indeed, as Batty argues, this way of thinking the city is not particularly novel:

> Analogies between the city and the human body, in which the city center is the heart that provides the motor to pump energy (in the form of traffic) . . . go back hundreds if not thousands of years, certainly to the time of Leonardo da Vinci and thence to the Greeks.[29]

Reaching a different conclusion, Richard Sennett argues that the development of the city is advanced alongside bodily knowledge in the Enlightenment. For instance, knowledge about circulation figured into city planning, as in L'Enfant's eighteenth-century plan for Washington, D.C., where the configuration of streets sought to create opportunities for mixing in the open air as a kind of urban lung.[30]

Movement is also a rich source for inspiration in quantified self-citynation, as animated cartographies of contemporary cities cinematically portray the liveliness of urban life. Jordan Crandall suggests that our measurements of movement "intoxicate us with the illusion of control, the ability to catalyze events and shape outcomes."[31] The portrayal of bike share programs in Paris and London mimic the metabolisms of an individual's check-ins on Foursquare. The percipients of these representations are made to focus on the clusters and dispersals, the rhythms of bodies and objects, set against the negative spaces of the map. In a more scholarly approach to movement, the global scale representations produced by the Urban Theory Lab, introduced earlier, elevate planetary movements as forms of extended urbanization, highlighting the territories that make urban life possible.

These fixations on movement and circulation can mask the inherit competition that this planning and visualization facilitates. Sennett reminds us of movement in the context of revolutionary Paris:

> The inability to reckon the urban crowd, or accept it whole, has of course to do with the people the crowd contained—people who were mostly poor. The poor, however, experienced movement in the city in

ways which lay beyond the scope of these prejudices. That experience crystallized in the meaning of market movements to the poor: the differences between survival and starvation they measured in the fluctuations of pennies or sous in the price of bread. The city's crowds wanted less market movement, more government regulation, fixity, and security. Physical movement in the city only sharpened their hunger pains.[32]

Movement facilitates competition, and systems that seek to measure, analyze, and represent movements set the condition for involvement in as well as interruption of the city. While unavoidable, perhaps, representations of movement serve both to highlight uneven development and provide the affective engine that aids in such continued fragmentations.

Movement of the city and movement of the body is not an inevitable function of daily life for everyone or every city. Of course, there is a political economy of movement. Figuring bodies and cities as organisms is to mask the very real injustices that are maintained through the calculative desires of quantification. Consider the rise of location-based services. LBSs have opened such opportunities for the multiscalar measurement of movement. These technologies, built into the hardware and software of mobile devices, enable an automated spatial curation, allowing individuals to experience the space-times they prefer, either knowing or unknowingly, through algorithms that render spatial decision making an opaque, hassle-free process. Certainly, privacy issues abound. In 2013 I received a letter, excerpted in Figure 23, from AT&T, my mobile carrier. These two policy changes by AT&T reflect a more general privatization of my movement, however announced as an address to questions of privacy: to "learn more about local consumers as a group" and creating location "characteristics" in order to connect advertisers with potential consumers, "more suited to your interests." These LBSs establish an attention marketplace, where the interruptive possibilities of mobile devices are up for sale. As my previous work on the rise of check-in services indicates, these systems can actually serve to narrow spatial habit and thereby interaction, creating grooves in the urban fabric for the already mobile.[33]

The quantification of proximity is a raw material to be extracted and refined. Competition in the provisioning of this digital spatial infrastructure forms around the possibilities for producing known habits, to alter behavior. From the perspective of mobile carriers like AT&T, movement patterns produce a topology of profit in the marketing of goods

> July 15, 2013
>
> Regarding Account Number: ...
>
> Dear Valued Customer,
>
> We know your privacy is important, so we've made it a priority to talk to you about it. We're revising our Privacy Policy to make it easier to understand, and we want to point out two new programs that could help us and other businesses serve you better.
>
> ...
>
> Program One: "External Marketing & Analytics Reports"
>
> ...
>
> For example, we might provide reports to retailers about the number of wireless devices in or near their store locations by time of day and day of week, together with the device users' collective information like ages and gender. This might allow a retailer in your neighborhood to learn more about local consumers as a group, but not about anyone individually, to improve its service.
>
> ...
>
> Program Two: Relevant Advertising including "Wireless Location Characteristics"
>
> ...
>
> We're currently creating a new "wireless location characteristic" that will help us use local geography as a factor in delivering ads. This doesn't mean you'll get more ads. It means that the ads that you do get from AT&T may be more suited to your interests.
>
> Location characteristics are types of locations – like "movie theaters." People who live in a particular geographic area might appear to be very interested in movies, thanks to collective information that shows wireless devices from that area are often located in the vicinity of movie theaters. We might create a "movies" characteristic for that area, and deliver movie ads to the people who live there.
>
> ...

FIGURE 23. A letter received in 2013 from AT&T describing new "characteristics" applied to user profiles.

and services. Indeed, from consumer location-based services to monitoring systems for urban transit, the question is: What kinds of habits might these sociotechnical systems reinforce and reward? As quantified self-city-nation is constituted on movement, as bodies and cities are brought into competition through the habituation of measurement and comparison, who—what bodies and what cities—will have the clearest advantage?

Fashion and Surveillance

Quantified self-city-nation utilizes software to make habitual the practices of measurement, analysis, representation—and behavioral adaptation. These technologies have been cast as surveillance by critical geographers, and further as sousveillance or the watching from below or within.[34] However, as these technologies become more about fashion and perception, perhaps it is helpful to think of them within the cultural form of the *selfie*.[35] Therefore, it is not sufficient to think of these techniques as only those that foster competition in the creation of profitable habits. For there is also self-satisfaction. To return to Amoore, these technologies may also "fulfil some of our desires and pleasures,"[36] and, as Crandall writes, in the pursuit of proprietary goals, these kinds of digital spatial techniques are used where "the event is not something to be prevented so much as courted."[37] The sharing of one's quantified self (as well as the cities we inhabit) is part of a competitive game of conspicuous mobility, fitness, and fashion.

Consider the proliferation of apps for the quantification of self. Personal activity monitors abound, inspiring us to know ourselves such that we can "live better."[38] Technologies like Automatic "help you make small changes in your driving habits that can lead to huge savings on gas over time," while another app called Spreadsheets allows you to diary your sex life, such that you can improve your record, including duration, days in a row, decibel peaks, and even thrusts per minute.[39] Family pets and even children can be fitted with sensors, providing caretakers with dynamic feeds, statistics, and digital maps.[40] These monitors attempt to quantify bodily knowledge to make it comparable and competitive, to utilize that knowledge to be bodied differently, and to showcase our bodily knowledge by sharing representations with our family, our friends, and our corporate providers.

The smart city operates similarly, to quantify the complexity of urban relations, to know how neighborhoods compare, to utilize that knowledge to be "smarter," and to showcase and compete among other cities by flexing the visualization muscle of an interoperable, habitual, and fashionable system. As is a frequented example among critical studies and popular commentary around the smart city, Rio's installation of a system with support from IBM and Bloomberg is more about perception than practice, more about being seen than seeing—although the practice of watching looms large. This is where quantified self-city-nation shows its cards, as a multiscalar system of attentional control. Biopower, the resource for making let live, blends into psychopower, as the constitution of care or the conditions around to what or whom is given attention.

While the question of the smart nation or secure homeland has been examined by political geographers,[41] I suggest that it borrows and trades on those metaphors that activate our thinking of smart bodies and cities.[42] Indeed, the American public had an opportunity to debate the renewal of the Patriot Act, which concentrated incredible power in the hands of the National Security Agency to collect and analyze our data, our movements, our attention. Its reform, however limited, became the USA Freedom Act. Know ourselves. Live better.

This nexus of fashion and surveillance highlights the shiftiness of bio- and psychopower, while suggesting a different register for the critique of quantification. As systems for "making countable," smart city infrastructure constitutes a new game board for competitive planetary urbanization—a process of peacocking that both attracts capital and appears to manage the risk of continued investment and development.[43] Calls for and lists of the most creative city have been eclipsed by lists of the smart*est* cities—no longer only a prioritization of educational institutions or percentages of persons holding advanced degrees but an advocacy of calculative efficiencies, predictive systems, and new technological capacities.[44] The use of location-based services furthers this high-stakes counting and centrally implicates the cartographer in these neuroses: selfies all the way down.

Infections

Of course, the risk of this critique is to treat quantification as a totalizing and pervasive assemblage without fracture or failure. The lines we draw indeed may draw us in, as this chapter has explored; however, there are always moments of breakage, a rest of the pen that enables pause, reconsideration, and the possibility for redirection. As infrastructure, systems of quantification are susceptible to disruptions.[45] As discourse, quantified self-city-nation is mere projection, a speculative proposal for a predictive location-aware future. How might we imagine and conjure opportunities for interrupting the quantified self-city-nation, as one form of quantification that reaches toxic levels? How might the all-too-easy slippage between personal activity monitors and smart city systems help us understand such opportunities?

To imagine the city as an organism is to also think the limits of such a city, just as there are limits to the body. Flesh *and stone,* as Sennett would insist. Organisms are vulnerable to infection, the cultivation of cells that react. I conclude then with a rash:

> We are aware that some of our customers have reported a skin irritation from wearing their Force device. We conduct testing in order to satisfy a variety of internationally accepted standards relating to the safety of the materials in our devices.[46]

In early 2014 Fitbit Force users went to social media to complain that their personal activity monitors, designed to be continuously worn, were causing rashes and even infections. Extensive testing and standards by Fitbit that were "internationally accepted" could not recreate the specific habits that put device to skin in ways that would leave a mark and cause irritation. Fitbit quickly discontinued the devices that were sold out in many stores across North America and eventually set about a process of a voluntary recall.

Although embarrassing and perhaps an isolated incident, these rashes speak to the intensive ways in which these technologies leave imprints—both physically and attentionally. In their organization of bodies, these technologies leave a mark and cause new habits, new relations to objects and others. Product recalls become trickier at the city scale, while the infections become more deadly. Rio's famed smart city system fuels the process of pacification in favelas, where militarized securitization is increasingly the norm. The clean forms represented in imagery from the central operations in Rio can be contrasted with the heterogeneity of form in favelas. The complexity of this city is but a proving ground for IBM's software; as Natasha Singer reports, "If the company can remake Rio as a smarter city, it can remake anywhere."[47] Still, psychopower meets biopower in the bloody incursions by Rio's police, as part of this remaking. Rashes, infections are abundant.

In these moments, of the routine and the bloody, the short term and the bodily, of quantification, of making countable and viewable, I return to the map as one such technology of representation. The quaint techniques that put pen to paper, vector to interactive display, are but part of a continuity of representational relations that increase in speed under quantified self-city-nation. The dualities outlined here become more pronounced, more rigid, more stratified. The digital map that guides us toward consumptive opportunities in our neighborhoods both creates and safeguards these neighborhoods.

Quantification, as an interoperable, proprietary system that fashions habits and surveils for the purposes of competition, is life lived under the spectacle, life as commodification. More than positivism then, quantification at these extremes is the process of imprinting. Here, I am reminded of Gunnar Olsson's encouragement of "a hermeneutics of suspicion, a mode of argument which says that the strength of cartographical reason lies less in its ability to tell the truth and more in its power to convince."[48] For Olsson and critical cartographers, the representation often hides the practices that give rise to its production: its sea monsters, its profit motives. We are left then with imprints, inscriptions, and affects.

Quantification is a force that attempts to imply its inevitability. Before the movement within space-time, and before any attention to or quantification of that movement, is a whole set of presuppositions that are largely ignored in the quantified self-city-nation. Paul Klee—the drawer, painter, and theoretician of lines in the early twentieth century—understood the importance of these presuppositions in the drawn line:

> Thus, analysis begins with the first element, the line, though, as Klee observes, it must be recognized that prior to the formal beginning, that is, before the first line is drawn, there lies an entire prehistory. This prehistory consists not only of the human longing for self-expression but also of a general condition of humankind that drives us on, by inner necessity, toward manifestation.[49]

What leads Klee to investigate the primal elements of the line is the investigation of that prehistory. This is precisely where the drawing, tracing, and mapping of new lines should bring us—toward that which conditions the line itself.

How are we convinced by quantified self-city-nation in its claims to truth? And how might we intervene in those conditions? As the smart city becomes the latest mantra for urban development, this logic for competition trickles down to lower-order cities. Civic hackers are flown from Silicon Valley to places like Lexington, Kentucky, to both open the data systems of municipal government and provide new private applications for these data networks and flows, new opportunities for profit. What are the various rashes caused by the imprint of these sociotechnical solutions that monitor ourselves, cities, and nations? Would we know how to see them develop and spread? New lines must both trace these imprints, as a record of their development, and map new interventions, new rhumb lines for alternative futures. Just as feminist economic geographers leveraged quantitative methods in the 1990s, to argue how method could be disentangled from epistemology, quantified self-city-nation must be redirected. Doing so would start from its status as a pharmacon for the formation of attention, for the enacting of a certain propriety that elevates a more modest openness and a culture of action that begins with sharing. Quantification is always crafted; it is not emergent and inevitable; points must be connected to form lines. As Klee recognized nearly a century ago, "The primal movement, the agent, is a point that sets itself in motion (genesis of form). A line comes into being."[50]

CHAPTER SIX

A Single Point Does Not Form a Line

> There is both a poetics *and* a politics of human geography and the two are closely connected.
>
> —DEREK GREGORY, "GREGORY D."

A single point does not form a line. To be certain, my thinking has excluded points and has privileged lines, not because the single point is too simple and beyond our discussion here, but rather that the single point remains too complex for my limited conceptualization. Points may have dimensionality, volume, tangency, even mass under the right conditions. They are coordinates, positions, markers, beginnings, ends, twists and turns, sites and sightings. However, the task ahead remains largely unchanged: how then to put points into motion? My suggestion: do not only draw and trace the lines of maps, but draw maps of traces. Responsive and responsible cartography is such an expression of care—to apprehend and elicit forces of thought and action. These maps of traces put single points into motion; they both are composed of lines and become lines. They take on mass, direction, and connection. They occupy poetics and politics.[1] They destratify while risking restratification. They attend. They move. Importantly, to the urgency of our current crises and our collective malaise, they enact an education and thereby traverse generations. I return to Bernard Stiegler, who beautifully narrates the significance of such care:

> Education is the fruit of the accumulated experience of generations. It develops a patina over time like the pebbles rolling in the current along the riverbed that they themselves constitute. Education is the transindividuation of individual memories engendered by individual experiences,

ones which, through being transmitted and developing a patina . . . have resulted in a collective memory constituted by the *attentional forms of knowledge: knowhow, lifeskills, cognitive and theoretical knowledges*.[2]

The movement of a line, as points put into motion, both illuminates complexities of the everyday and challenges our most deeply felt convictions expressed in thought and action. For Stiegler, "a philosophy of care assumes a philosophy of attention, especially in our epoch where an 'attention economy' dominates."[3] I suggest new lines as both a record and a proposal, a memory and a pedagogy, to capture our desires and craft our care. The map, imbricated by such lines in our location-aware society, is an artifact that links science and culture, industry and academe, government and governance. However, it is its becoming an event that most animates my thinking.

The task of this text was to trace these lines and explore their ends both imagined and realized, across five fractures that divide and connect, create ruptures and link breakages: criticality, digitality, movement, attention, and quantification. There are other troubles, to be sure. Instead of offering a philosophy that supersedes the weight of ideas past, I have brought these theories into reconsideration. Additive, not subtractive. Indeed, while Arthur Robinson's *Elements of Cartography* dominated mid- to late twentieth-century thought regarding the map and provided the primary foundation for contemporary map design research, I suggest that the force of this thought simply does not take maps seriously enough. He writes, "Most maps are functional in that they are designed, like a bridge or a house, for a purpose."[4] This is hardly the full story, and yet Alan MacEachren's *How Maps Work* in 1995 also does not greatly alter the imagination of the cartographer, nor the register within which the cartographer may ask questions of their craft.[5]

Instead, as Tom Conley writes, "The event of the map becomes a virtual passage from a perception of detail to infinity that moves with and through projective folds."[6] Put differently, the lines of a map are rambunctious, both opening and closing and bringing together yet-unwritten histories and yet-unrealized futures. The previous five chapters are my modest and partial attempts at exploratory essays for recognizing the thickening of the lines of criticality, digitality, movement, attention, and quantification, in the moments of their fracture. In other words, if these read strange to those within the traditions and conventions of GIScience and GIS & Society scholarship, then the text has been productive. I am not attempting to assert the critical grounds on which digital methods in mapping tackle movement, create and attend to an audience, through unending quantification of sociospatial relations. There is much more to be mapped in these traces, not in the sense of new territories to be represented but to intervene with new force. In other words, the previous five lines and their fracturing enables a discussion of the role and limits of criticality, the restratifying tendencies that surround origin stories of digitality, the productive elusiveness of movement in mapping, the attentive conditions of the drawn and traced line, and the restlessness of quantification as it takes new forms. Points in motion. And still more questions: Why the urgency? I return to education as intergenerational stratification.

Sun sets. Sun rises. It is a fantastic cosmology that the earth seemingly holds still while the sun moves across the sky. Geocentrism is perhaps one of our more basic humanistic vices—to assume a stable position amid the destabilizing forces and movements all around us. More than vice, it is a primary observation that marks time, produces habit, and configures the rhythms of work and play. Science asks us to reset these accumulated inscriptions, to more abstractly vision a different model of interaction, to put into motion that which feels stationary and to stabilize that which constitutes the measurement of life itself. Indeed, our very simple notions of progress spring forward, dawn to dusk, in series of periodicities of days, months, seasons, and years that we later assemble and call history. These individual experiences, to witness the sun moving across the sky, become individual memory. We attend to the world in this most basic way, to bear witness to time, to keep hope that the sun will return again—if only we pay attention.

It is in the context of this attentional practice, to attend to what matters, that I read a survey of Americans conducted by the National Science Foundation and released in 2014 regarding our knowledge of the relationship between the sun and the earth.[7] Indeed, one in four Americans seems to believe that the sun orbits the earth. A prominent geographer who earned his bachelor of arts degree at Harvard in 1905, Isaiah Bowman noted that these kinds of limited observations would continue to linger. In 1934 he wrote as director of the American Geographical Society:

> When prejudice is fortified by limited experience it may linger for centuries. There are millions of persons who still believe the earth to be flat because it is not perceptibly curved as one looks at it casually. It is fantastic, a defiance of common sense and a violation of experience to say that the sun does not move across the sky when it so obviously does. It requires wider experience, more general observation than can be made in one spot, and the application of technical methods and measurements to *prove* the rotation of the round earth.[8]

More experience, more observation, more knowledge. More attention. Indeed, more care. Education is such a system of care. It is a curiosity then that in just a decade following this statement Bowman would reportedly assure the president of Harvard, Jim Conant, of his decision to close the geography program—a decision that would change the face of geographic education in North America.[9] Instead, geography would remain and grow largely as a project of public institutions of higher education. Care matters; care materializes.

The sun stretches slightly farther across the sky. And one of four may believe what seems so plainly seen. How do we respond to this NSF report on our general knowledge? Do we laugh? Do we lower our heads in shame? Inasmuch as this survey will likely motivate advocates of science education to take up their microscopes, telescopes, and models, I would like to take a perhaps less obvious rejoinder—that which highlights the contiguity of relations that make resilient *this* or *that* knowledge. For me, this fact about our collective knowledge, that one in four Americans believes the sun orbits the earth, is more an indictment of our collective memory—and we are all culpable. As the report indicates, science coverage makes up a quite small percentage of traditional media—less than 2 percent—so it is perhaps unsurprising that many more Americans may know the current controversies of Justin Bieber than of the relationship between our planet and the sun.

We are all culpable, because to what we attend, because to what we pay attention matters and materializes. Education as such a system of care is increasingly in a losing position, as the cultural industry produces information, an enduring resource, while our capacity to pay attention is incredibly finite. A location-aware society prefigures this capacity and, further, speeds up the effects of their finitude. Under such a society of the spectacle, Guy Debord writes during the unrest of the 1960s, that our individual and collective consciousness has become targeted, and while religious contemplation was perhaps an earliest expression of this targeting, the spectacle now operates to frame to what we pay attention.[10] Everyday life becomes the raw material of click bait.

That one in four Americans believes the sun orbits the earth does not only mark the ignorance of our fellow citizens, but instead highlights our collective indulgence in and prioritization of systems that actually cause us *not* to care. We simply change the channel; we scroll further in our social media feeds; we distract ourselves. Debord understood the urgency of this collective malaise in 1955. He writes:

> We need to flood the market—even if for the moment merely the intellectual market—with a mass of desires whose fulfillment is not beyond

the capacity of humanity's present means of action on the material world, but only beyond the capacity of the old social organization.[11]

The market of ideas needed then, as now, to be saturated by a "mass of desires," not merely those drives which have so toxified humanity's facility to take care. While Stiegler proposes an economy of contribution, as a culture of sharing that might reestablish the circuitry for collective and intergenerational memory, I suggest to slow the process of mapmaking, to engage in slow mapping. While not a panacea, slow maps enable different habits, alternative forms of care.[12]

Slow Maps

At the recent annual meetings of the North American Cartographic Information Society there has been a curious sentiment among organizers that maintains a division between practical cartographers and academic cartography, and a not so subtle sense among some attendees that "if you cannot do, teach." Eyes roll as academic cartographers discuss interaction principles just as drool collects at the corners of attendees' mouths as they gaze longingly at the latest map candy on display. At the 2015 meetings in Minneapolis, the usual awards given to cartographic submissions were split into two categories: a designation for a map displaying notable research and one for a map displaying notable design. This of course begs two obvious questions: Should not a well-designed map also be well researched? Should a well-researched map not necessarily be well designed? Indeed, the calibrations of the field seem askew.

Instead, what might a map intervention look like that slows the process of map communication, to directly counter map design research that seeks to speed up the process of map reading? Not only to put points into motion, but put points into motion *well*. I propose that slow maps value the intense concentration and investment of attention necessary to understand and learn about geographical relations. Ephemeral observations and experiences beget ephemeral knowledge. Difficult observation and hard-fought experience begets lasting understanding, perhaps. The slow map intensifies and produces desire. As maps of traces, slow maps resist the temptation of a fast-map, neo-Robinsonian functionalism, in order to foster different thought and action. The map, through such slow map experimentation, becomes theory in the sense of a creative conceptualization.

To champion such a deliberate slowing of the map event in these contemporary space-times is challenging. When questions around the availability and necessity of the Internet were being asked within the offices of National Public Radio in 1994, a memorandum was issued: "If you do not want to use Internet, simply do nothing."[13] This reasonable recommendation seems quaint to us now, as the pervasiveness of the Internet causes us to hardly recall a time in which one used the Internet, as one might use a telephone. Instead many of us are continuously connected, materially, imaginatively, and discursively—giving rise to my insistence for "a more complex understanding of the embrace of technology."[14] To think the map as an event is to introduce a series of pedagogical moments, to motivate a mapping that is beyond effectivity and toward affectivity. This move already introduces agents that slow the process, to not skip over "essentially subjective" aspects of the map,[15] but to elaborate and magnify them. The five troubles introduced here are but anticipated breakages in the rhizomes created by the cartographic. These breakages collude and resist distinction. Neither criticality nor digitality solely binds cartographic thought and action. Movement-attention-quantification occur as differences of quality along a continuity of relations: new lines in the location-aware future. Make slow maps. Tinker and fail.

To examine a piece of hand-drawn cartography, like Raisz's *Landforms*, is to seemingly witness the movement of the cartographer's pen, the force of their body against the manuscript. These are the maps that compel our attention, to attend to the movements of their lines amid the static. We trace them with fingertips and eyes until they come to ends. We are moved by these lines, affectively, just as we move about through their instruction, effectively. Beyond appreciation, their grasp is elemental, a most basic correspondence between craft and utility. Phenomenologically, then, all lines are new. While we can return to these maps, their static lines betray their movements, their stirrings.

In this text I have fixated on the concept and gritty materiality of the drawn and traced line. However, while these new lines are discrete at times and blurry at others, sometimes thick and encompassing and at other times thin and discriminating, their interpretations are also productively ambiguous if resistive. In a brief passage of *Cinema 1,* Deleuze relatedly argues why attention to the specific technical and artistic capacities of film enables a whole philosophy. On the topic of the line, he writes:

> It links up man and woman and the cosmos. It connects desires, suffering, errors, trials, triumphs, appeasements. It connects moments of intensity, as so many points through which it passes. It connects the living and the dead.[16]

Details, to infinity. These lines are not "the encompassing stroke of a great contour, but the broken stroke of a line of the universe, across the holes."[17]

Similarly, the lines of our maps are much more than the expressions of technical facility and expertise, more than the poetic artistry that relates ground to graphic and back to ground again, more than the behaviors of maker and user. New lines are additive and ambitious. It is not our task to only ever draw or trace them, but our unique role to force them toward different ends and, in doing so, take responsibility over all that they describe and all that they nudge from the margins into the bounding box of our collective conviction. Utilizable and relevant geographic representation. Yes, we *do* GIS.

ACKNOWLEDGMENTS

This project began as a nudge in 2010. I am so thankful to those who have helped me think my work in a different register, to grant me some space and time to make a mess of fields and domains, to imagine alternative audiences and lean on different voices. Throughout, Sarah Elwood has helped me carry this forward and assisted me in recognizing the long rhythms of academic life as they get translated into the productive distractions of everyday thought and action. Sam Kinsley similarly encouraged me to think more generally about my work and helped me think through the implications of Bernard Stiegler. Sonya Prasertong and Jessa Loomis were also brilliant cointerviewers with community partners, helping bring to ground a series of uneasy gaps between theory and practice. Students at Ball State, Kentucky, and Harvard have helped push along these ideas, especially Eric Huntley, Amber Bosse, Dan Cockayne, Ryan Cooper, Eric Nost, Bryan Preston, Andrea Craft, Jessi Breen, Zulaikha Ayub, Conor O'Shea, Chris Alton, and Dan Koff. Kentucky Geography is such a remarkable place and I am so fortunate to have the best colleagues, who have punctuated the work with beer cheese and porch drinks, especially Patricia Ehrkamp, Matt Zook, Jeremy Crampton, Rich Donohue, Anna Secor, Rich Schein, Sue Roberts, Betsy Beymer-Farris, Carolyn Finney, and David Nemer. I will continue to pinch myself.

I also have amazing coconspirators and friends in the discipline, who have fostered this work and helped me to imagine its implications, including Nadine Schuurman, James Ash, Britta Ricker, Wen Lin, Agnieskza Leszczynski, Taylor Shelton, Mark Graham, Rebecca Lave, Neil Brenner, Paul Simpson, Vicky Lawson, Michael Brown, Jim Thatcher, Jin-Kyu Jung, Meghan Cope, Rob Kitchin, Mike Goodchild, Rachel Pain, David O'Sullivan, Eric Sheppard, Trevor Barnes, Barbara Meloni, Robin McElheny, Wendy

Guan, and so many others: thank you. An early version of the Introduction was originally invited by Nadine. Chapter 2 benefited from supportive audiences at Harvard University, Penn State University, Syracuse University, University at Buffalo, UC–Berkeley, and the University of Illinois, as well as the opportunity to work with Trevor for our coauthored article in *Big Data and Society*. Chapter 3 benefited from discussions with participants in the Deleuze Reading Group at Kentucky, including Jeff Peters, Curtis Pomilia, Eric Huntley, Jeremy Crampton, and Kendra Sanders.

Many heartfelt thanks to my family and friends who put so many points into motion for me, including my parents, Kay and Norman; my brother, Dallas; and my grandparents, Charles, Anna, Joy, and Dean; my close friends on long journeys and short, Adam, Travis, and Matt D.; Jon F. and Katie F.; Jon L., Conor, and Aneesha; Andy, Mike, and Katie K.; Erica, Jon G., and Sara G.; David, Jason, and Freddie; Jordan and Anna; Ben, Greg, Jim, Austin, Corey, and Josie. From Pumpkin Center to Maryville, from Seattle to Muncie, Barcelona, Cambridge, and Lexington, I so adore our collective memories and look forward to future trips.

Finally, to Jason Weidemann at the University of Minnesota Press, my sincere appreciation for helping guide the project from its earliest coffee-stained proposals to its culminating microbrews.

NOTES

Preface

1. J. Brian Harley, "Deconstructing the Map," *Cartographica* 26 (1989): 1–20. A reading of Harley's correspondence in the British Library would indicate that he knew this article would stir trouble, and specifically among more traditional cartographers and advocates of mapmaking in the discipline of geography.

2. See Matthew W. Wilson, "Towards a Genealogy of Qualitative GIS," in *Qualitative GIS: A Mixed Methods Approach*, edited by Meghan Cope and Sarah A. Elwood (London: Sage, 2009), 156–70; Nadine Schuurman and Geraldine Pratt, "Care of the Subject: Feminism and Critiques of GIS," *Gender, Place and Culture* 9, no. 3 (2002): 291–99. Generally speaking, Bruno Latour's "Why Has Critique Run Out of Steam? From Matters of Fact to Matters of Concern," *Critical Inquiry* 30 (Winter 2004): 225–48, similarly problematizes critique within the Left.

3. Haraway suggests a kind of constructivism that is "about contingency and specificity but not epistemological relativism" in *Modest_Witness@Second_Millennium. Femaleman©_Meets_Oncomouse™: Feminism and Technoscience* (New York: Routledge, 1997), 99. She brings this forward in *When Species Meet* (Minneapolis: University of Minnesota Press, 2008); see also "*When Species Meet*: Staying with the Trouble," *Environment and Planning D: Society and Space* 28 (2010): 53–55; *Staying with the Trouble: Making Kin in the Chthulucene* (Durham, N.C.: Duke University Press, 2016).

4. Gilles Deleuze and Félix Guattari, *A Thousand Plateaus: Capitalism and Schizophrenia* (Minneapolis: University of Minnesota Press, 1987), 25.

5. Here, I use *technoculture* to emphasize the often-entangled relationship between technology and culture, borrowing the term from Constance Penley and Andrew Ross, *Technoculture* (Minneapolis: University of Minnesota Press, 1991). They write in the introduction to this edited collection: "Wary, on the one hand, of the disempowering habit of demonizing technology as a satanic mill of

domination, and weary, on the other, of postmodernist celebrations of the technological sublime, we selected contributors whose critical knowledge might help to provide a realistic assessment of the politics—the dangers *and* the possibilities—that are currently at stake in those cultural practices touched by advanced technology" (xii).

6. Nicholas R. Chrisman, "What Does 'GIS' Mean?," *Transactions in GIS* 3, no. 2 (1999): 175.

7. Chrisman, "Design of Geographic Information Systems Based on Social and Cultural Goals," *Photogrammetric Engineering & Remote Sensing* 53, no. 10 (1987): 1367–70.

8. Chrisman attributes this to K. J. Dueker and D. Kjerne in "What Does 'GIS' Mean?" (178). Indeed, this was the definition of GIS I was taught in my undergraduate course by Dr. Mark Corson.

9. Henri Lefebvre, *The Production of Space* (Cambridge, Mass.: Blackwell, 1991), 124–25.

10. Ibid., 126.

11. Gunnar Olsson, *Lines of Power/Limits of Language* (Minneapolis: University of Minnesota Press, 1991), 181.

Introduction

1. The ugliness of the rapid reconfiguration of value in the academy can be easily witnessed in the for-profit proliferation of impact measures, across the United Kingdom in various configurations of assessment frameworks and exercises and especially pronounced in the United States with websites like Google Scholar, Academic.edu, and ResearchGate, multiplying the effects of quantification of individual scholars. Indeed, in personal conversation in late 2015 with a software developer at Academic.edu, I learned that nearly a third of their several million users signed into the site primarily to check the analytics of their scholarship—who was viewing them, from what countries, and so on: Narsacism.edu.

2. See Jim Thatcher, Luke Bergmann, Britta Ricker, Reuben Rose-Redwood, David O'Sullivan, Trevor J. Barnes, Luke R. Barnesmoore, et al., "Revisiting Critical GIS," *Environment and Planning A* 48, no. 5 (2016): 815–24.

3. See Nadine Schuurman's contribution, "Is the Rubric 'Critical GIScience' Effective? An Argument for Theoretical GIScience," in Matthew W. Wilson, Barbara S. Poore, Francis Harvey, Mei-Po Kwan, David O'Sullivan, Marianna Pavlovskaya, Nadine Schuurman, and Eric Sheppard, "Theory, Practice, and History in Critical GIS: Reports on an AAG Panel Session," *Cartographica* 44, no. 1 (2009): 5–16.

4. Neil Smith, "History and Philosophy of Geography: Real Wars, Theory Wars," *Progress in Human Geography* 16 (1992): 257–71; Joel Wainwright,

Geopiracy: Oaxaca, Militant Empiricism, and Geographical Thought (New York: Palgrave, 2013).

5. For instance, see Agnieszka Leszczynski, "Spatial Media/tion," *Progress in Human Geography* 39, no. 6 (2015): 729–51.

6. Constance Penley and Andrew Ross, *Technoculture* (Minneapolis: University of Minnesota Press, 1991). See also footnote 5 in the preface.

7. Indeed, there is much skepticism these days about social theoretical development in cartography and GIScience, as if debates and discussions regarding concepts and implications have already happened and require no further innovation.

8. Gunnar Olsson, *Lines of Power/Limits of Language* (Minneapolis: University of Minnesota Press, 1991), 180–81.

9. Note Matthew Sparke's critique of that vision, in "Escaping the Herbarium: A Critique of Gunnar Olsson's 'Chaism of Thought-and-Action,'" *Environment and Planning D: Society and Space* 12 (1994): 207–20; Matthew Sparke, "The Return of the Same in Geography: A Reply to Olsson," *Environment and Planning D: Society and Space* 12 (1994): 226–28. See also Gunnar Olsson, *Abysmal: A Critique of Cartographic Reason* (Chicago: University of Chicago Press, 2007).

10. Indeed, consider their discussions on ontology and GIScience ontologies: Agnieszka Leszczynski, "Poststructuralism and GIS: Is There a 'Disconnect'?," *Environment and Planning D: Society and Space* 27 (2009): 581–602; Jeremy W. Crampton, "Being Ontological: Response to 'Poststructuralism and GIS: Is There a "Disconnect"?,'" *Environment and Planning D: Society and Space* 27 (2009): 603–8. Leszczynski offers a rebuttal, "This attention to 'knowing'—how it is that we come to know the world as technolog*ical*—attests to both an ontology and epistemology of GIScience technologies," in "Rematerializing GIScience," *Environment and Planning D: Society and Space* 27 (2009): 611.

11. Olsson, *Lines of Power*, 190.

12. Here, I am indebted to the work of Meghan Cope and Sarah Elwood, for resisting the all-too-easy designation of qualitative GIS as practice or method. Instead, "practice" and "method" are largely unsuitable for such mapping work; see Meghan Cope and Sarah Elwood, "Conclusion: For Qualitative GIS," in *Qualitative GIS: A Mixed Methods Approach*, edited by Meghan Cope and Sarah A. Elwood (London: Sage, 2009), 171–77. Although, also see Paul Harrison, "In the Absence of Practice," *Environment and Planning D: Society and Space* 27, no. 6 (2009): 987–1009.

13. Gilles Deleuze and Félix Guattari, *A Thousand Plateaus: Capitalism and Schizophrenia* (Minneapolis: University of Minnesota Press, 1987).

14. Ibid., 21; this sentiment is what animated my thinking in "Map the Trace," *ACME: An International E-Journal for Critical Geographies* 13, no. 4 (2014): 583–85.

15. Deleuze and Guattari, *A Thousand Plateaus*, 21.

16. Ibid.

17. This was largely my point in my invited response to Manuel B. Aalbers, "Do Maps Make Geography? Part 1: Redlining, Planned Shrinkage, and the Places of Decline," *ACME: An International E-Journal for Critical Geographies* 13, no. 4 (2014): 525–56; see Wilson, "Map the Trace."

18. Olsson, *Lines of Power,* 181.

19. I return to this notion of drawers who trace and tracers who draw as a kind of technopositionality, of both the capacity to draw lines and make maps and the capacity to trace lines and maps and understand the importance/implications of these graphic representations. To further complicate, I also refer to these drawn/ traced lines as both metaphorical and literal, as lines of direction and connection (what might be thought as discourse) and lines of graphic representation.

20. Donna Jeanne Haraway, "The Promises of Monsters: A Regenerative Politics for Inappropriate/d Others," in *Cultural Studies,* edited by Lawrence Grossberg, Cary Nelson, and Paula A. Treichler (New York: Routledge, 1992), 295.

21. Ibid.

22. See Nadine Schuurman, "Critical GIS: Theorizing an Emerging Science," *Cartographica* 36, no. 4 (1999): 7–108.

23. John Pickles, ed., *Ground Truth: The Social Implications of Geographic Information Systems* (New York: Guilford, 1995).

24. John Pickles, "Ground Truth 1995–2005," *Transactions in GIS* 10, no. 5 (2006): 765.

25. Tom Poiker, "Preface," *Cartography and Geographic Information Systems* 22, no. 1 (1995): 3–4; Pickles, *Ground Truth;* Eric Sheppard, "GIS and Society: Towards a Research Agenda," *Cartography and Geographic Information Systems* 22, no. 1 (1995): 5–16.

26. National Center for Geographic Information and Analysis, "NCGIA Research Initiatives," http://www.ncgia.ucsb.edu/research/initiatives.html.

27. See Matthew W. Wilson, "Towards a Genealogy of Qualitative GIS," in *Qualitative GIS: A Mixed Methods Approach,* edited by Meghan Cope and Sarah A. Elwood (London: Sage, 2009), 156–70.

28. Rebecca Lave, Matthew W. Wilson, Elizabeth S. Barron, Christine Biermann, Mark A. Carey, Chris S. Duvall, Leigh Johnson, et al., "Intervention: Critical Physical Geography," *Canadian Geographer* 58, no. 1 (2014): 1–10.

29. See special issues on volunteered geographic information and neogeography: Matthew W. Wilson and Mark Graham, "Situating Neogeography," *Environment and Planning A* 45, no. 1 (2013): 3–9; Agnieszka Leszczynski and Matthew W. Wilson, "Theorizing the Geoweb," *GeoJournal* 78, no. 6 (July 2013): 915–19.

30. See chapter 4, and recent work on attention economies, including Elvin K. Wyly, "The City of Cognitive-Cultural Capitalism," *City* 17, no. 3 (2013): 387–94; Sam Kinsley, "Memory Programmes: The Industrial Retention of Collective Life,"

cultural geographies 22, no. 1 (2015): 155–75; Bernard Stiegler, "Relational Ecology and the Digital Pharmakon," *Culture Machine* 13 (2012): 1–19.

31. Nadine Schuurman and Mei-Po Kwan, "Guest Editorial: Taking a Walk on the Social Side of GIS," *Cartographica* 39, no. 1 (Spring 2004): 1–3; David O'Sullivan, "Geographical Information Science: Critical GIS," *Progress in Human Geography* 30, no. 6 (2006): 783–91; see also Eric Sheppard's contribution in Matthew W. Wilson, Barbara S. Poore, Francis Harvey, Mei-Po Kwan, David O'Sullivan, Marianna Pavlovskaya, Nadine Schuurman, and Eric Sheppard, "Theory, Practice, and History in Critical GIS: Reports on an AAG Panel Session," *Cartographica* 44, no. 1 (2009): 5–16.

32. See debate in Dawn Wright, Michael F. Goodchild, and James D. Proctor, "GIS: Tool or Science? Demystifying the Persistent Ambiguity of GIS as 'Tool' versus 'Science,'" *Annals of the Association of American Geographers* 87, no. 2 (1997): 346–62; and John Pickles, "Tool or Science? GIS, Technoscience, and the Theoretical Turn," *Annals of the Association of American Geographers* 87, no. 2 (June 1997): 363–72.

33. See danah boyd and Kate Crawford, "Critical Questions for Big Data: Provocations for a Cultural, Technological, and Scholarly Phenomenon," *Information, Communication & Society* 15, no. 5 (2012): 662–79; Matthew W. Wilson, "Morgan Freeman Is Dead and Other Big Data Stories," *cultural geographies* 22, no. 2 (2015): 345–49; Craig Dalton and Jim Thatcher, "What Does a Critical Data Studies Look Like, and Why Do We Care? Seven Points for a Critical Approach to 'Big Data,'" *Environment and Planning D: Society and Space* (May 2014), http://societyandspace.org/2014/05/12/what-does-a-critical-data-studies-look-like-and-why-do-we-care-craig-dalton-and-jim-thatcher/.

34. Gilles Deleuze and Félix Guattari, *On the Line* (New York: Semiotext(e), 1983), 18.

35. Olsson, *Lines of Power*, 181.

36. Deleuze and Guattari, *On the Line*, 10. Here, I am also encouraged by Tim Ingold in his writing on blobs and lines; although I disagree on his distinction: "Blobs have volume, mass, density; they give us materials. Lines have none of these. What they have, which blobs do not, is torsion, flexion and vivacity. They give us life"; see Tim Ingold, *The Life of Lines* (London: Routledge, 2015), 4. In my reading, lines do take on mass; they materialize.

37. Gilles Deleuze, *Cinema 1: The Movement-Image*, translated by Hugh Tomlinson and Barbara Habberjam (London: Bloomsbury Academic, 2013), 35.

38. See Wilson, "Map the Trace."

39. Smith, "Real Wars."

40. Harley, "Deconstructing the Map," 1.

41. Indeed, Pickles's *A History of Spaces* could be considered an elaboration of Harley's project, a point that Leszczynski has made in "Poststructuralism and

GIS," 591; John Pickles, *A History of Spaces: Cartographic Reason, Mapping, and the Geo-Coded World* (New York: Routledge, 2004).

42. Arthur Howard Robinson, *The Look of Maps: An Examination of Cartographic Design* (Redlands, Calif.: Esri, 2010), 73.

43. See Matthew W. Wilson, "'Training the Eye': Formation of the Geocoding Subject," *Social and Cultural Geography* 12, no. 4 (June 2011): 357–76; Matthew W. Wilson, "Data Matter(s): Legitimacy, Coding, and Qualifications-of-Life," *Environment and Planning D: Society and Space* 29, no. 5 (2011): 857–72.

44. Rebecca Krinke, "Unseen/Seen: The Mapping of Joy and Pain," last modified 2011, http://www.rebeccakrinke.com/Projects/Unseen-Seen-The-Mapping-of-Joy-and-Pain.

45. Guy Debord, "Introduction to a Critique of Urban Geography," in *Critical Geographies: A Collection of Readings*, edited by Harald Bauder and Salvatore Engel-Di Mauro (Kelowna, BC: Praxis, 2008), 23–27.

46. Deleuze and Guattari, *On the Line*, 95.

47. Nadine Schuurman, "Trouble in the Heartland: GIS and Its Critics in the 1990s," *Progress in Human Geography* 24, no. 4 (2000): 569–90.

48. Susan Roberts, "Realizing Critical Geographies of the University," *Antipode* 32, no. 3 (2000): 230–44; Harald Bauder, "Learning to Become a Geographer: Reproduction and Transformation in Academia," *Antipode* 38, no. 4 (2006): 671–79; Bonnie Kaserman and Matthew W. Wilson, "On Not Wanting It to Count: Reading Together as Resistance," *Area* 41, no. 1 (2009): 26–33.

49. Greg Miller, "Meet the Man Who Wants to Teach the World to Make Maps," *Wired*, July 16, 2013, http://www.wired.com/wiredscience/2013/07/anthony-robinson-mooc/.

50. See this debate played out in the blog and comments of a recent geography PhD turned spatial consultant, "Beware the Social Theorists in Geography," last modified August 21, 2015, http://www.justinholman.com/2015/08/21/beware-the-social-theorists-in-geography/, as compared with Stan Openshaw, "A View on the GIS Crisis in Geography, or, Using GIS to Put Humpty-Dumpty Back Together Again," *Environment and Planning A* 23, no. 5 (1991): 621–28; and Stan Openshaw, "Further Thoughts on Geography and GIS: A Reply," *Environment and Planning A* 24, no. 4 (April 1992): 463–66.

51. Jacques Steinberg, "SAT's Reality TV Essay Stumps Some," *New York Times*, March 16, 2011.

52. Smith, "Real Wars."

53. Neil Smith, "Neo-Critical Geography, or, the Flat Pluralist World of Business Class," *Antipode* 37, no. 5 (2005): 889.

54. This echoes similar skepticism by Pickles, "Tool or Science?"; Eric Sheppard, "Branding GIS: What's 'Critical'?," *Cartographica* 44, no. 1 (2009): 13–16; and O'Sullivan, "Geographical Information Science."

1. Criticality

1. Bruno Latour, "Why Has Critique Run Out of Steam? From Matters of Fact to Matters of Concern," *Critical Inquiry* 30 (Winter 2004): 225–48.

2. For instance, Paul Harrison considers the outside of criticality (as affirmation); see Paul Harrison, "After Affirmation, or, Being a Loser: On Vitalism, Sacrifice, and Cinders," *GeoHumanities* 1, no. 2 (2015): 285–306.

3. Jeff Pruchnic, "Postcritical Theory? Demanding the Possible," *Criticism* 54, no. 4 (2012): 637–57.

4. Guy Debord, "Introduction to a Critique of Urban Geography," in *Critical Geographies: A Collection of Readings*, edited by Harald Bauder and Salvatore Engel-Di Mauro (Kelowna, BC: Praxis, 2008), 26.

5. William Warntz, "Preface to 'Fred K. Schaefer and the Science of Geography,'" *Harvard Papers in Theoretical Geography: Special Papers Series no. A*, November 1, 1968, i–ii.

6. Jouni Häkli quoted in Guntram H. Herb, Jouni Häkli, Mark W. Corson, Nicole Mellow, Sebastian Cobarrubias, and Maribel Casas-Cortes, "Intervention: Mapping Is Critical!," *Political Geography* 28 (August 2009): 335.

7. Pruchnic, "Postcritical Theory?," 639.

8. Donna Jeanne Haraway, *Modest_Witness@Second_Millennium. FemaleMan©_Meets_Oncomouse™: Feminism and Technoscience* (New York: Routledge, 1997), 255.

9. Many thanks to Sarah Elwood for this phrasing—as we contemplated writing a different chapter but felt ensnared by the Friday Harbor narrative; see Matthew W. Wilson and Sarah A. Elwood, "Capturing," in *The Sage Handbook of Human Geography,* edited by Roger Lee, Noel Castree, Rob Kitchin, Victoria Lawson, Anssi Paasi, Chris Philo, Sarah Radcliffe, Susan Roberts, and Charles Withers (London: Sage, 2014), 235–53.

10. Nadine Schuurman, "Critical GIS: Theorizing an Emerging Science," *Cartographica* 36, no. 4 (1999): 7–108. Schuurman discussed with me the provenance of this curious opportunity, to publish her entire PhD dissertation. Following her PhD defense, her committee took her to lunch at the University of British Columbia faculty club. At that time, Brian Klinkenberg (UBC professor) was the editor of the journal, and he plainly asked Nadine if she would consider putting her whole dissertation into the journal. Undoubtedly, the immediate availability of this material was like a shot in the arm for the fledgling subfield.

11. See Nadine Schuurman and Geraldine Pratt, "Care of the Subject: Feminism and Critiques of GIS," *Gender, Place and Culture* 9, no. 3 (2002): 291–99.

12. Schuurman, "Critical GIS," 9, original emphasis.

13. Matthew W. Wilson, "Towards a Genealogy of Qualitative GIS," in *Qualitative GIS: A Mixed Methods Approach,* edited by Meghan Cope and Sarah A. Elwood (London: Sage, 2009), 164–66. Here, I defined *technopositionality* as "a positionality in conducting research that is simultaneously about and with the

technology. It is 'techno' in the sense that its relationship with technology is hybrid—a taking up of the discourses and technicalities of the machine" (164).

14. See Nadine Schuurman, "Trouble in the Heartland: GIS and Its Critics in the 1990s," *Progress in Human Geography* 24, no. 4 (2000): 569–90.

15. Peter J. Taylor, "GKS," *Political Geography Quarterly* 9 (July 1990): 212.

16. Stan Openshaw, "A View on the GIS Crisis in Geography, or, Using GIS to Put Humpty-Dumpty Back Together Again," *Environment and Planning A* 23, no. 5 (1991): 623.

17. Peter J. Taylor and Michael Overton, "Further Thoughts on Geography and GIS," *Environment and Planning A* 23, no. 8 (1991): 1087–90.

18. Stan Openshaw, "Further Thoughts on Geography and GIS: A Reply," *Environment and Planning A* 24, no. 4 (April 1992): 464.

19. Neil Smith, "History and Philosophy of Geography: Real Wars, Theory Wars," *Progress in Human Geography* 16 (1992): 257–71.

20. Eric Sheppard, "Automated Geography: What Kind of Geography for What Kind of Society?," *Professional Geographer* 45, no. 4 (1993): 460.

21. John Pickles, ed., *Ground Truth: The Social Implications of Geographic Information Systems* (New York: Guilford, 1995); Tom Poiker, "Preface," *Cartography and Geographic Information Systems* 22, no. 1 (1995): 3–4.

22. Trevor M. Harris, Daniel Weiner, Timothy A. Warner, and Richard Levin, "Pursuing Social Goals through Participatory Geographic Information Systems," in *Ground Truth: The Social Implications of Geographic Information Systems*, edited by John Pickles (New York: Guilford Press, 1995), 196–222.

23. William J. Craig, "The Internet Aids Community Participation in the Planning Process," *Computers, Environment and Urban Systems* 22, no. 4 (1998): 393–404.

24. See the debates surrounding the Bowman Expeditions, as discussed in a 2010 special issue of a prominent geography journal: Philip E. Steinberg, "Professional Ethics and the Politics of Geographic Knowledge," *Political Geography* 29 (2010): 413.

25. Eric Sheppard, "Knowledge Production through Critical GIS: Genealogy and Prospects," *Cartographica* 40, no. 4 (2005): 5–21.

26. Here, I'm reminded of my friend and colleague Agnieszka Leszczynski's tweet on September 24, 2015: "Can we please stop talking about 'Critical GIS'? We need to stop talking about Critical GIS" (https://twitter.com/agaleszczynski/status/647160781746757632). Her insistent question, part tongue-in-cheek, is: the technology has changed, so why has our agenda of critique not also changed?

27. See Matthew W. Wilson and Monica Stephens, "GIS as Media?," in *Mediated Geographies and Geographies of Media*, edited by Susan Mains, Julie Cupples, and Chris Lukinbeal (New York: Springer, 2015), 209–21; Agnieszka Leszczynski, "Spatial Media/tion," *Progress in Human Geography* 39, no. 6 (2015): 729–51.

28. See Sara Fabrikant's entry into this historical fascination: Sara Irina Fabrikant, "Commentary on 'A History of Twentieth-Century American Academic Cartography' by Robert McMaster and Susanna McMaster," *Cartography and Geographic Information Science* 30, no. 1 (2003): 81–84.

29. James Corner, "The Agency of Mapping: Speculation, Critique and Invention," in *Mappings,* edited by Denis E. Cosgrove (London: Reaktion Books, 1999), 213–52.

30. For an overview of the conceptualization of the map within geography, see Rob Kitchin, Chris Perkins, and Martin Dodge, "Thinking about Maps," in *Rethinking Maps: New Frontiers in Cartographic Theory,* edited by Martin Dodge, Rob Kitchin, and Chris Perkins (London: Routledge, 2009), 1–25.

31. See the report on basic science knowledge and the public understanding of the relationship between the earth and the sun in the United States by the National Science Foundation, *Science and Engineering Indicators,* edited by National Center for Science and Engineering Statistics (Arlington: National Science Foundation, 2014), http://www.nsf.gov/statistics/seind14/index.cfm/chapter-7.

32. See Bernard Stiegler, "Relational Ecology and the Digital Pharmakon," *Culture Machine* 13 (2012): 1–19; Sam Kinsley, "Memory Programmes: The Industrial Retention of Collective Life," *cultural geographies* 22, no. 1 (2015): 155–75; James Ash, "Attention, Videogames and the Retentional Economies of Affective Amplication," *Theory, Culture & Society* 29, no. 6 (2012): 3–26; Patrick Crogan and Sam Kinsley, "Paying Attention: Towards a Critique of the Attention Economy," *Culture Machine* 13 (2012): 2–29.

33. While I explore this in more detail in chapter 4, "Attention," this way of thinking of the work of the map enables an alternative framing of the urgency of geographic representation.

34. Corner, "Agency of Mapping," 216.

35. I contemplate this in a recent piece called "Morgan Freeman Is Dead and Other Big Data Stories," of course playing off the fact that Morgan Freeman's death quickly became a globally trending topic on Twitter—the only problem being that Freeman was still very much alive. This should give us pause as we consider the design potential for big data—these data sets are not purely evidence of phenomena but are phenomena in and of themselves.

36. See Matthew W. Wilson and Mark Graham, "Neogeography and Volunteered Geographic Information: A Conversation with Michael Goodchild and Andrew Turner," *Environment and Planning A* 45, no. 1 (2013): 10–18.

37. See Matthew W. Wilson, "On the Criticality of Mapping Practices: Geodesign as Critical GIS?," *Landscape and Urban Planning* 142 (2015): 226–34.

38. "Geodesign," Esri, http://www.esri.com/products/arcgis-capabilities/geodesign/overview.

39. Time geographies and visualizations of space-time prisms have long vexed quantitative geographers; see Mei-Po Kwan, "Feminist Visualization: Re-Envisioning

GIS as a Method in Feminist Geographic Research," *Annals of the Association of American Geographers* 92, no. 4 (2002): 645–61.

40. Doreen Massey, *For Space* (London: Sage, 2005).

41. See important interventions in spatial history, such as Anne Kelly Knowles and Amy Hillier, *Placing History: How Maps, Spatial Data, and GIS Are Changing Historical Scholarship* (Redlands, Calif.: Esri, 2008). Hayles takes up the Stanford Spatial History Project (http://spatialhistory.stanford.edu) counterposing Massey's "dream" of coevalness.

42. N. Katherine Hayles, *How We Think: Digital Media and Contemporary Technogenesis* (Chicago: University of Chicago Press, 2012), 198.

43. For further discussion of the role of movement in mapping, as told through a discussion of animated cartography, see chapter 3, "Movement."

44. Scott Bell and Maureen Reed, "Adapting to the Machine: Integrating GIS into Qualitative Research," *Cartographica* 39, no. 1 (Spring 2004): 55–66; Samuel F. Dennis Jr., "Prospects for Qualitative GIS at the Intersection of Youth Development and Participatory Urban Planning," *Environment and Planning A* 38 (2006): 2039–54; LaDona Knigge and Meghan Cope, "Grounded Visualization: Integrating the Analysis of Qualitative and Quantitative Data through Grounded Theory and Visualization," *Environment and Planning A* 38 (2006): 2021–37; Mei-Po Kwan and LaDona Knigge, "Doing Qualitative Research Using GIS: An Oxymoronic Endeavor?," *Environment and Planning A* 38 (2006): 1999–2002; Marianna Pavlovskaya, "Non-Quantitative GIS," in *Qualitative GIS: A Mixed-Methods Approach,* edited by Meghan Cope and Sarah A. Elwood (London: Sage, 2009), 13–37; Wilson, "Towards a Genealogy of Qualitative GIS."

45. For an example of this development, see Jin-Kyu Jung, "Software-Level Integration of CAQDAS and GIS," in *Qualitative GIS: A Mixed-Methods Approach,* edited by Sarah A. Elwood and Meghan Cope (London: Sage, 2009), 115–36.

46. Enrique Chagoya, *Le Cannibale Moderniste,* mixed media on paper on linen (1999).

47. Neil Brenner, "Theses on Urbanization," *Public Culture* 25, no. 1 (2013): 85–114.

48. I introduced the notion of technopositionality in my discussion of the philosophy—as in the concepts created—by qualitative GIS as a form of critical GIS in Wilson, "Towards a Genealogy of Qualitative GIS."

49. Here, I am reminded that these kinds of curiosities can cause more technically oriented eyes to roll; see Rich Donohue, "On the Political Economy of the GeoJSON Format," 2014, http://medium.com/latitude-beta/on-the-political-economy-of-the-geojson-format-8e7f38b9f5d8. Here, Donohue takes up my question—I am the "Dr. X" to whom he refers—regarding the provenance of GeoJSON.

50. David DiBiase, Michael DeMers, Ann Johnson, Karen Kemp, Ann Taylor Luck, Brandon Plewe, and Elizabeth Wentz, *Geographic Information Science and Technology Body of Knowledge* (Washington, D.C.: UCGIS and the Association of American Geographers, 2006).

51. Bruno Latour, *Science in Action: How to Follow Scientists and Engineers through Society* (Cambridge, Mass.: Harvard University Press, 1987), 227. He continues, "All these objects occupy the beginning and the end of a similar accumulation cycle; no matter whether they are far or near, infinitely big or small, infinitely old or young, they all end up at such scale that a few men or women can dominate them by sight; at one point or another, they all take the shape of a flat surface of paper that can be archived, pinned on a wall and combined with others; they all help to reverse the balance of forces between those who master and those who are mastered."

52. And earlier with regard to participatory mapping, see Michael Stone, "Map or Be Mapped," *Whole Earth* (Fall 1998): 54–55.

53. See Wilson and Elwood, "Capturing," where we discuss the ways in which mapping, in a variety of forms, engages the notion of capture.

54. Joe Bryan, "Force Multipliers: Geography, Militarism, and the Bowman Expeditions," *Political Geography* 29 (2010): 414–16; Melquiades (Kiado) Cruz, "A Living Space: The Relationships between Land and Property in the Community," *Political Geography* 29 (2010): 420–21; Peter H. Herlihy, "Self-Appointed Gatekeepers Attack the American Geographical Society's First Bowman Expedition," *Political Geography* 29 (2010): 417–19; Philip E. Steinberg, "Professional Ethics and the Politics of Geographic Knowledge," *Political Geography* 29 (2010): 413.

55. See Neil Smith, *American Empire: Roosevelt's Geographer and the Prelude to Globalization* (Berkeley: University of California Press, 2003); see also chapter 2 of this volume, "Digitality," where Bowman is more directly weaved into the story of the digital map.

56. Erwin Raisz, "Orthoapsidal World Maps," *Geographical Review* 33, no. 1 (January 1943): 132–34.

57. Raisz authored what is considered the first cartography textbook in English: Erwin Raisz, *General Cartography*, McGraw-Hill Series in Geography, 1st ed. (New York: McGraw-Hill, 1938).

58. Howard T. Fisher, *Mapping Information: The Graphic Display of Quantitative Information* (Cambridge, Mass.: Abt Associates, 1982), 26–27.

59. These connections are discussed further in chapter 2, "Digitality."

60. William Bunge, "Theoretical Geography" (PhD diss., University of Washington, 1960), 26–27.

61. Susan Schulten, "Richard Edes Harrison Reinvented Mapmaking for World War 2 Americans," *New Republic,* May 20, 2014.

62. "Perspective Map: Harrison Atlas Gives Fresh New Look to Old World," *Life,* February 28, 1944, 56–61. This feature in *Life* highlights Harrison's disdain, "Harrison is always up in arms against the academic cartographers who stubbornly refuse to break away from conventions. He is also thoroughly annoyed at careless Americans who refuse to see the facts that maps show them."

63. Ibid., 61.

2. Digitality

1. William Bunge, "Fred K. Schaefer and the Science of Geography," *Harvard Papers in Theoretical Geography: Special Papers Series*, no. A, November 1, 1968, 1–22.

2. Schaefer had attempted a critique of Richard Hartshorne, the author of *The Nature of Geography*, in 1939 by establishing the role of methodology; Fred K. Schaefer, "Exceptionalism in Geography: A Methodological Examination," *Annals of the Association of American Geographers* 43, no. 3 (September 1953): 226–49. He opens this manuscript with, "The methodology of a field is not a grab bag of special techniques. In geography such techniques as map making, 'methods' of teaching, or historical accounts of the development of the field are still often mistaken for methodology."

3. I am grateful to Nick Chrisman, a former mentor of mine at the University of Washington, and his autobiographical account of the LCGSA: Nicholas R. Chrisman, *Charting the Unknown: How Computer Mapping at Harvard Became GIS* (Redlands, Calif.: ESRI, 2006).

4. Here, we could lump the efforts of the GIS & Society movement, outlined in chapter 1, with the technology sections of major periodicals, including *Wired* and the *New York Times*.

5. Esri, "Disaster Response Program," www.esri.com/services/disaster-response.

6. See Susan Roberts and Richard H. Schein, "Earth Shattering: Global Imagery and GIS," in *Ground Truth: The Social Implications of Geographic Information Systems*, edited by John Pickles (New York: Guilford, 1995), 171–95; Matthew W. Wilson, "Continuous Connectivity, Handheld Computers, and Mobile Spatial Knowledge," *Environment and Planning D: Society and Space* 32, no. 3 (2014): 535–55.

7. Donna Jeanne Haraway, *Modest_Witness@Second_Millennium. Femaleman©_Meets_Oncomouse™: Feminism and Technoscience* (New York: Routledge, 1997), 151.

8. Erwin Josephus Raisz, *General Cartography* (New York: McGraw-Hill, 1938), vii.

9. Will C. van den Hoonaard, *Map Worlds: A History of Women in Cartography* (Waterloo, Ontario: Wilfrid Laurier University Press, 2013); Judith Tyner, "Millie the Mapper and Beyond: The Role of Women in Cartography since World War II," *Meridian* 15 (1999): 23–28; Marianna Pavlovskaya and Kevin St. Martin, "Feminism and Geographic Information Systems: From a Missing Object to a Mapping Subject," *Geography Compass* 1, no. 3 (2007): 583–606.

10. Paul N. Edwards, *The Closed World: Computers and the Politics of Discourse in Cold War America* (Cambridge, Mass.: MIT Press, 1996); Trevor J. Barnes, "Lives Lived and Lives Told: Biographies of Geography's Quantitative Revolution," *Environment and Planning D: Society and Space* 19 (2001): 409–29; Trevor J. Barnes, "Geography's Underworld: The Military-Industrial Complex,

Mathematical Modelling and the Quantitative Revolution," *Geoforum* 39 (2008): 3–16; Trevor J. Barnes and Matthew W. Wilson, "Big Data, Social Physics, and Spatial Analysis: The Early Years," *Big Data & Society* 1, no. 1 (April–June 2014): 1–14.

11. Erwin Raisz, "Geography at Harvard," *Science* 115, no. 2989 (April 1952): 405–6; Neil Smith, "'Academic War over the Field of Geography': The Elimination of Geography at Harvard, 1947–1951," *Annals of the Association of American Geographers* 77, no. 2 (1987): 155–72; Neil Smith, *American Empire: Roosevelt's Geographer and the Prelude to Globalization* (Berkeley: University of California Press, 2003); Saul B. Cohen, "Reflections on the Elimination of Geography at Harvard, 1947–51," *Annals of the Association of American Geographers* 78, no. 1 (March 1988): 148–51; Michael S. DeVivo, *Leadership in American Academic Geography: The Twentieth Century* (Lanham, Md.: Lexington Books, 2015); William Bunge, *Fitzgerald: Geography of a Revolution* (Cambridge, Mass.: Schenkman, 1971); Nik Heynen, "Marginalia of a Revolution: Naming Popular Ethnography through William W. Bunge's Fitzgerald," *Social and Cultural Geography* 14, no. 7 (2013): 744–51.

12. Freshmen Seminar Activity, 1966, Box 18, Problem Statement [Student Problem Sets] Folder, Papers of Howard T. Fisher, HUGFP 62.7, Harvard University Archives, Cambridge, Mass.

13. Ibid.

14. For example, see the use of automation at the USGS; Patrick H. McHaffie, "Towards the Automated Map Factory: Early Automation at the U.S. Geological Survey," *Cartography and Geographic Information Science* 29, no. 3 (2002): 193–206.

15. It is shameful that very little has been written or documented about Betty Benson (born October 6, 1924, and died October 15, 2008), one of many women enrolled in the work of the LCGSA in the 1960s and 1970s. Her obituary notes that she held a master's degree in anthropology from Northwestern and worked in Evanston as a computer programmer; see "Death Notice: Betty Tufvander Benson," *Chicago Tribune,* October 19, 2008.

16. Transcription from Northwestern University Conference, 1970, Box 15, Miscellaneous Correspondence, Memoranda and Other Papers 2 Folder, Papers of Howard T. Fisher, HUGFP 62.7, Harvard University Archives, Cambridge, Mass.

17. A reproduction of this agreement, dated June 18, 1930, is available for viewing at the Recording Secretary's Office of Alumni Affairs and Development at Harvard University, although the agreement predates the creation of the Recording Secretary's Office.

18. Raisz had worked with Davis's student Douglas Johnson at Columbia University, while completing his doctorate in geology. At the Institute at Harvard, Raisz was named lecturer in cartography. See Arthur Howard Robinson, "Erwin

Josephus Raisz, 1893–1968," *Annals of the Association of American Geographers* 60, no. 1 (1970): 189–93.

19. Raisz, *General Cartography*.

20. This is a photo of the sixth edition of the original manuscript, available for viewing at the Harvard Map Collection. Note the lighter shade where Raisz added Lake Mead. The size of this reservoir would have changed from the original manuscript in 1939 to this revision in 1952.

21. See also the Raisz experiments with the "armadillo projection," Erwin Raisz, "Orthoapsidal World Maps," *Geographical Review* 33, no. 1 (January 1943): 132–34.

22. Brooke E. Marston and Bernhard Jenny, "Improving the Representation of Major Landforms in Analytical Relief Shading," *International Journal of Geographic Information Science* 29, no. 7 (2015): 1144–65.

23. Smith, "Academic War"; see also DeVivo, *Leadership*.

24. Actually, a department of geography was never fully established at Harvard. Conant's remarks ended an ongoing discussion about the creation of a department.

25. Robinson details in Erwin Raisz's obituary that following the demise of the institute, Raisz held teaching assignments at the University of Virginia, University of Florida, University of British Columbia, and Clark University; Robinson, "Erwin Josephus Raisz." See Edward A. Ackerman, "Derwent Stainthrope Whittlesey," *Geographical Review* 47, no. 3 (July 1957): 443–45; "Geography at Harvard," *Harvard Crimson*, November 26, 1956; "Well-Known Geographer Derwent Whittlesey Dies," *Harvard Crimson,* November 26, 1956.

26. Personal communication with Dick Morrill, February 7, 2014; J. Nicholas Entrikin and Stanley D. Brunn, *Reflections on Richard Hartshorne's The Nature of Geography, Occasional Publications of the Association of American Geographers* (Washington, D.C.: Association of American Geographers, 1989), 76–80.

27. See Barnes, "Lives Lived and Lives Told"; Richard L. Morrill, "Recollections of the 'Quantitative Revolution's' Early Years: The University of Washington 1955–65," in *Recollections of a Revolution*, edited by M. Billinge (London: Macmillan, 1984), 57–72.

28. William Bunge, "Theoretical Geography" (PhD diss., University of Washington, 1960).

29. Ibid., 24.

30. Ibid., 26, emphasis original.

31. Ibid., 28–29.

32. Waldo R. Tobler, "Automation and Cartography," *Geographical Review* 49, no. 4 (October 1959): 526–34.

33. Spokane Blight by Census Block, Box 11, Accession Number 3365-87-13, Papers of Ed Horwood, Special Collections, University of Washington Libraries, Seattle, Wash.

NOTES TO CHAPTER 2

34. The SYMAP User's Reference Manual, fifth edition, second printing (October 1975), contains a two-page history of the program, establishing the key break from Horwood's Card Mapping Program, with three objectives: first, "to create maps which would be complete in themselves"; second, "to display spatial data graphically with variable darkness and texture"; and third, "to provide a method of interpolating irregularly spaced data values." Fisher corresponded with Horwood in October and November 1975 on the topic of crediting Horwood's workshop in August 1963. In addition to Betty Benson (at Northwestern), other programmers mentioned included those at Harvard: Robert Russell, Donald Shepard, Marion Manos, and Kathleen Reine. See SYMAP User's Reference Manual, pp. ii–iii, Box 22, Papers of Howard T. Fisher, HUGFP 62.7, Harvard University Archives, Cambridge, Mass.

35. Press Release from Northwestern, December 9, 1964, Box 16, Northwestern University Folder, Papers of Howard T. Fisher, HUGFP 62.7, Harvard University Archives, Cambridge, Mass.

36. SYMAP User's Reference Manual, Section 3, p. 10, Box 22, Papers of Howard T. Fisher, HUGFP 62.7, Harvard University Archives, Cambridge, Mass.

37. Howard Fisher to David K. Camfield, May 11, 1966, Box 20, Southern Illinois University Folder, Papers of Howard T. Fisher, HUGFP 62.7, Harvard University Archives, Cambridge, Mass.

38. The goals for the lab were part of a broader agenda by the mid-1970s to transform the LCGSA into a Laboratory for Computer Technology that would be a support unit for the Graduate School of Design; see Goals of the Lab, Box 15, Miscellaneous Correspondence, Memoranda and Other Papers 2 Folder, Papers of Howard T. Fisher, HUGFP 62.7, Harvard University Archives, Cambridge, Mass.

39. Schmidt was working with Richard Duke at MSU, also with funding from Ford. Schmidt recalls meeting Fisher at an American Institute of Planning conference. At the time Schmidt was working in Louisville, Kentucky, where he completed the SYMAP correspondence course.

40. In some sense, Harvard's Center for Geographic Analysis (CGA), formed in the fall of 2006, is the latest incarnation of a similar set of arrangements—a methodologically focused center without a disciplinary home.

41. Howard Fisher to Claire Neikirk, July 25, 1966, Box 12, LCG Library Folder, Papers of Howard T. Fisher, HUGFP 62.7, Harvard University Archives, Cambridge, Mass.

42. Waldo Tobler to Howard Fisher, September 20, 1966, Box 22, Tobler Waldo University of Michigan Folder, Papers of Howard T. Fisher, HUGFP 62.7, Harvard University Archives, Cambridge, Mass.

43. Howard Fisher to Arthur Robinson, 1966, Box 23, University of Wisconsin Folder, Papers of Howard T. Fisher, HUGFP 62.7, Harvard University Archives, Cambridge, Mass.

44. Fisher submitted a portfolio in November 1973 for an American Institute of Architects Fellowship. Computer mapping was included, along with his work on prefabricated houses, shopping centers, and curtain wall systems. See SYMAP Example, Box 18, Portfolio for AIA Fellowship Nomination 1 Folder, Papers of Howard T. Fisher, HUGFP 62.7, Harvard University Archives, Cambridge, Mass.

45. Transcription from Northwestern University Conference, 1970, Box 15, Miscellaneous Correspondence, Memoranda and Other Papers Folder 2, Papers of Howard T. Fisher, HUGFP 62.7, Harvard University Archives, Cambridge, Mass.

46. Robert Williams to Howard Fisher, March 22, 1966, Box 23, Williams Robert Folder, Papers of Howard T. Fisher, HUGFP 62.7, Harvard University Archives, Cambridge, Mass.

47. Howard Fisher to Robert Williams, April 21, 1966, Box 23, Williams Robert Folder, Papers of Howard T. Fisher, HUGFP 62.7, Harvard University Archives, Cambridge, Mass.

48. Williams to Fisher, March 22, 1966.

49. Howard Fisher to John Bland, May 13, 1966, Box 15, N-Correspondence Folder, Papers of Howard T. Fisher, HUGFP 62.7, Harvard University Archives, Cambridge, Mass.

50. Howard Fisher to Eric Teicholz, August 7, 1967, Box 22, T-Correspondence Folder, Papers of Howard T. Fisher, HUGFP 62.7, Harvard University Archives, Cambridge, Mass.

51. Howard Fisher to William L. Knight, August 28, 1967, Box 14, Memos to Allan Schmidt Folder 1, Papers of Howard T. Fisher, HUGFP 62.7, Harvard University Archives, Cambridge, Mass.

52. Introduction to SYMAP Correspondence Course, ca. 1967, Box 21, SYMAP Correspondence Course Materials Folder 1, Papers of Howard T. Fisher, HUGFP 62.7, Harvard University Archives, Cambridge, Mass.

53. Summary of Introductory Correspondence Course, ca. 1967, Box 21, SYMAP Correspondence Course Materials Folder 1, Papers of Howard T. Fisher, HUGFP 62.7, Harvard University Archives, Cambridge, Mass.

54. Proposed 1968–69 Freshmen Seminar, Box 15, Miscellaneous Correspondence, Memoranda and Other Papers Folder 2, Papers of Howard T. Fisher, HUGFP 62.7, Harvard University Archives, Cambridge, Mass.

55. Fisher's correspondence with Ted Wilcox, director of Harvard's freshmen seminar program, would indicate that the seminar was not offered by Harvard College. However, my discussions with lab alumni Geoff Dutton complicate this matter slightly (as he recalls that particular seminar). Correspondence with Edward T. Wilcox, 1968, Box 23, W-Correspondence Folder, Papers of Howard T. Fisher, HUGFP 62.7, Harvard University Archives, Cambridge, Mass.

56. The Ford Foundation would eventually agree to an additional $40,000 to fund Fisher after 1970, toward the completion of this book. (Of course, the book would not be completed until 1982, published after Fisher's death in 1979.)

57. G. M. Gaits, "Thematic Mapping by Computer," *Cartographic Journal* 6, no. 1 (June 1969): 50–68.

58. Howard Fisher to Donald S. Shepard, June 16, 1970, Box 20, Shepard Donald S. Folder, Papers of Howard T. Fisher, HUGFP 62.7, Harvard University Archives, Cambridge, Mass.

59. Geoff Dutton, Memoir, May 5, 1966, prepared for the occasion of the dedication of the Harvard Center for Geographic Analysis, shared with the author.

60. Notes on the book project, Box 20, Significance, Scope, and Organization Folder, Papers of Howard T. Fisher, HUGFP 62.7, Harvard University Archives, Cambridge, Mass.

61. "Computers Can Draw Maps, and Public Works Men Can Use Them," *APWA Reporter*, July 1970, 21, Box 15, Miscellaneous Correspondence, Memoranda and Other Papers Folder 1, Papers of Howard T. Fisher, HUGFP 62.7, Harvard University Archives, Cambridge, Mass.

62. Howard Fisher to Local Board No. 1, June 10, 1969, Box 9, Goodrich John C. Folder, Papers of Howard T. Fisher, HUGFP 62.7, Harvard University Archives, Cambridge, Mass.

63. Notes on the book project, Box 20, Significance, Scope, and Organization Folder, Papers of Howard T. Fisher, HUGFP 62.7, Harvard University Archives, Cambridge, Mass.

64. Tom Thayer is a central figure in a discussion of the Hamlet Evaluation System (HES); see Stathis N. Kalvas and Matthew Adam Kocher, "The Dynamics of Violence in Vietnam: An Analysis of the Hamlet Evaluation System (HES)," *Journal of Peace Research* 46, no. 3 (2009): 335–55; Thomas C. Thayer, *War without Fronts: The American Experience in Vietnam, Westview Special Studies in Military Affairs* (Boulder, Colo.: Westview, 1985). For the single, conspicuous piece of correspondence with Thayer in the Fisher papers, see Thomas C. Thayer to Howard Fisher, March 1, 1967, Box 22, T-Correspondence Folder, Papers of Howard T. Fisher, HUGFP 62.7, Harvard University Archives, Cambridge, Mass. Here, Thayer indicates that Robert Taylor at the Advanced Research Projects Agency (ARPA, now DARPA) would be brought in to discuss the funding of Fisher's research. Taylor was tasked to set up a computer at the Military Assistance Command Vietnam (MACV); see Robert Taylor, "An Interview with Robert Taylor," by William Aspray, *The Center for the History of Information Processing,* February 28, 1989, Charles Babbage Institute, University of Minnesota. My own conversations with Carl Steinitz indicated that members of the lab were certainly contacted regarding computer mapping techniques in support of the Department of Defense efforts in Vietnam.

65. Here, I am grateful to archival research by Oliver Belcher, "Data and Difference: Race, Violence, and the Making of the Hamlet Evaluation System in Vietnam" (paper presented at the Annual Meeting of the Association of American Geographers, San Francisco, March 29, 2016).

66. Transcription from Northwestern University Conference, 1970, Box 15, Miscellaneous Correspondence, Memoranda and Other Papers Folder 2, Papers of Howard T. Fisher, HUGFP 62.7, Harvard University Archives, Cambridge, Mass.

67. Notes on the book project, Box 20, Significance, Scope, and Organization Folder, Papers of Howard T. Fisher, HUGFP 62.7, Harvard University Archives, Cambridge, Mass.

68. Howard Fisher to Carl Steinitz, March 17, 1967, Box 20, Steinitz Carl F. Folder, Papers of Howard T. Fisher, HUGFP 62.7, Harvard University Archives, Cambridge, Mass.

69. Howard Fisher to Ed Horwood, October 13, 1975, Box 14, Misc. Correspondence 1975–1977 Folder 1, Papers of Howard T. Fisher, HUGFP 62.7, Harvard University Archives, Cambridge, Mass.

70. Howard Fisher to Allan Schmidt, December 9, 1975, Box 14, Misc. Correspondence 1975–1977 Folder 2, Papers of Howard T. Fisher, HUGFP 62.7, Harvard University Archives, Cambridge, Mass.

71. Howard Fisher to William Warntz, August 10, 1971, Box 23, W-Correspondence Folder, Papers of Howard T. Fisher, HUGFP 62.7, Harvard University Archives, Cambridge, Mass.

72. The lab would eventually become embroiled in issues of software royalties and the use of the Harvard brand on lab materials. See Chrisman, *Charting the Unknown*.

73. Fisher, *Mapping Information*, xix.

74. Ibid., 10.

75. Mark S. Monmonier, "Review of Mapping Information: The Graphic Display of Quantitative Information by Howard T. Fisher," *Annals of the Association of American Geographers* 74, no. 2 (June 1984): 348–51; Mark S. Monmonier, *How to Lie with Maps* (Chicago: University of Chicago Press, 1991).

76. Howard Fisher to Robert Williams, April 29, 1968, Box 23, Yale University Folder, Papers of Howard T. Fisher, HUGFP 62.7, Harvard University Archives, Cambridge, Mass.

3. Movement

1. Walter Isard, "On Notions and Models of Time," *Papers of the Regional Science Association* 25 (1970): 8; Gilles Deleuze, *Cinema 2: The Time-Image*, translated by Hugh Tomlinson and Robert Galeta (New York: Bloomsbury Academic, 2013), 262.

2. I overview this concern with time and liveliness in chapter 1 of this volume, "Criticality," as conjured by the development of spatial history (and the spatial humanities, more generally). This chapter can be considered a further scratching of this particular itch, by taking up another earlier development: animated cartography.

3. However, ethics in GIScience (unit GS6) is considered a core unit, while temporal phenomena (unit DM5) is not; see David DiBiase, Michael DeMers, Ann Johnson, Karen Kemp, Ann Taylor Luck, Brandon Plewe, and Elizabeth Wentz, *Geographic Information Science and Technology Body of Knowledge* (Washington, D.C.: UCGIS and the Association of American Geographers, 2006), 29.

4. Matthew W. Wilson, "Towards a Genealogy of Qualitative GIS," in *Qualitative GIS: A Mixed Methods Approach*, edited by Meghan Cope and Sarah A. Elwood (London: Sage, 2009), 156–70; see also chapter 1 of this volume, "Criticality."

5. Foucault defines *dispositif*, or apparatus, as "the said as much as the unsaid" in his interview in "The Confessions of the Flesh," in *Power/Knowledge: Selected Interviews and Other Writings 1972–1977*, edited by Colin Gordon (New York: Pantheon Books, 1977), 194–95.

6. André Pierre Colombat, "Deleuze and Signs," in *Deleuze and Literature*, edited by John Marks and Ian Buchanan (Edinburgh: Edinburgh University Press, 2000), 29.

7. Henri Bergson, *Matter and Memory* (New York: Zone Books, 1988).

8. Gilles Deleuze, *Cinema 1: The Movement-Image*, translated by Hugh Tomlinson and Barbara Habberjam (New York: Bloomsbury Academic, 2013), 194–95.

9. Allan Schmidt, *A Pictorial History of the Expansion of the Metropolitan Area* (Lansing: Michigan State University, 1967). Schmidt described, in personal communication with the author, that after completing the SYMAP correspondence course program, he was hired by Richard Duke at Michigan State as assistant director of the METRO project, funded by Ford. Schmidt wrote to Fisher to request a copy of the SYMAP software, and a programmer converted it from running on an IBM 7094 to the CDC 3600 at MSU. Here, he motivated a group of graduate students to record every subdivision at the state land registry, to map the coordinates of these subdivisions as individual snapshots for the eventual animated map.

10. Brian Fung, "Watch Twitter Explode along with Ferguson," *Washington Post*, August 14, 2014, www.washingtonpost.com/blogs/the-switch/wp/2014/08/14/watch-twitter-explode-along-with-ferguson; see also the larger version of the map hosted by Carto, http://srogers.cartodb.com/viz/4a5eb582-23ed-11e4-bd6b-0e230854a1cb/embed_map.

11. As Robinson points out in the Foreword of *The Look of Maps*, his writing was critiqued and evaluated by important figures of early twentieth-century geography and cartography, including J. K. Wright and Richard Edes Harrison; see Arthur H. Robinson, *The Look of Maps: An Examination of Cartographic Design* (Madison: University of Wisconsin Press, 1952), xi–xiii.

12. Judith Tyner, "Elements of Cartography: Tracing Fifty Years of Academic Cartography," *Cartographic Perspectives* 151 (Spring 2005): 4–13.

13. Erwin Josephus Raisz, *General Cartography*, 2nd ed. (New York: McGraw-Hill, 1948).
14. Robinson, *The Look of Maps*, xi.
15. Ibid., 4.
16. Ibid., 10.
17. Ibid., 19.
18. For a discussion of the encounters with Robinson and George Jenks at the 1970 conference on quantitative mapmaking at Northwestern, see chapter 2.
19. Much of *The Look of Maps* centers on the importance, style, and employment of lettering. Those who may skim through these chapters, thinking they are of a vintage quality, may miss the subtle ways in which the techniques of evaluation of lettering in advertising is privileged by Robinson in his advocacy of functional design in mapmaking.
20. Matt Edney argues that Robinson's approach was an internal history of cartography; see Matthew H. Edney, "Putting 'Cartography' into the History of Cartography: Arthur H. Robinson, David Woodward, and the Creation of a Discipline," *Cartographic Perspectives* 51 (Spring 2005): 14–29.
21. Jacques Bertin, *Semiology of Graphics* (Madison: University of Wisconsin Press, 1983); Robert E. Roth, "Cartographic Interaction Primitives: Framework and Synthesis," *Cartographic Journal* 49, no. 4 (November 2012): 376–95.
22. Robinson, *The Look of Maps*, 73.
23. Jeremy W. Crampton, "Maps as Social Constructions: Power, Communication and Visualization," *Progress in Human Geography* 25, no. 2 (2001): 235–52.
24. Doreen Massey, *For Space* (London: Sage, 2005), 13.
25. Ibid., 23.
26. Henri Bergson, *Matter and Memory* (New York: Zone Books, 1988); Deleuze, *Cinema 1;* Deleuze, *Cinema 2.*
27. Massey, *For Space*, 24.
28. Ibid.
29. Ibid., 30.
30. Kenneth Field, June 17, 2014, http://twitter.com/kennethfield/status/478775510386741248.
31. Daniel Dorling, "Stretching Space and Splicing Time: From Cartographic Animation to Interactive Visualization," *Cartography and Geographic Information Systems* 19, no. 4 (1992): 215.
32. Deleuze, *Cinema 1,* 2.
33. Ibid., 5.
34. Ibid., 6.
35. I am not suggesting that animated cartography is an example of what Deleuze terms the affection-image, but rather that his conceptualization of the affection-image

in film provides some interesting, and perhaps more specific, techniques of analysis in our understanding of maps that move. One could also interrogate animated maps using the perception-image or action-image. These are beyond my analysis here but may provide yet further ways to more specifically map the traces created by these moving maps.

36. Deleuze, *Cinema 1*, 90–91. Note that Deleuze equates the close-up with the face; both are affects. The face or close-up could be an object, such as a clock face, or a human face, or a condition of an object itself.

37. Ibid., 102. Many thanks to my fellow participants in the *Cinema 1* reading group, especially Jeff Peters, for discussing the importance of this passage with me.

38. Ibid., 17.

39. Norman J. W. Thrower, "Animated Cartography," *Professional Geographer* 11, no. 6 (1959): 10.

40. These federal funds ignited controversy over the inclusion of loyalty oaths, amid McCarthyism. After protest, and Kennedy's election, these oaths were removed as a requirement.

41. Thrower, "Animated Cartography," 12.

42. Bruce Cornwell and Arthur Howard Robinson, "Possibilities for Computer Animated Films in Cartography," *Cartographic Journal* 3, no. 2 (1966): 79–82.

43. Bertin, *Semiology of Graphics*, 42.

44. Waldo R. Tobler, "A Computer Movie Simulating Urban Growth in the Detroit Region," *Economic Geography* 46 (June 1970): 234–40.

45. Ibid., 236.

46. Harold Moellering, "The Potential Uses of a Computer Animated Film in the Analysis of Geographical Patterns of Traffic Crashes," *Accident Analysis and Prevention* 8, no. 4 (1976): 217.

47. I share in this academic genealogy. My adviser at the University of Washington, Tim Nyerges, read for his PhD with Moellering at Ohio State (graduating in 1980). While Nyerges's work does not emphasize the movement of the map, per se, his interest in the ways in which map and map interactions can impact decision making and planning is approached from a similar perspective: that the perception of a map is something that can be controlled for in behavioral experiments with maps. See Timothy L. Nyerges, "Geographic Information Abstractions: Conceptual Clarity for Geographic Modeling," *Environment and Planning A* 23 (1991): 1483–99; Timothy L. Nyerges, "Analytical Map Use," *Cartography and Geographic Information Systems* 18, no. 1 (January 1991): 11–22; Timothy L. Nyerges, "How Do People Use Geographical Information Systems?," in *Human Factors in Geographical Information Systems*, edited by David Medyckyj-Scott and Hilary Hearnshaw (London: Belhaven Press, 1993), 37–50.

48. George F. McCleary Jr., "Discovering Cartography as a Behavioral Science," *Journal of Environmental Psychology* 7 (1987): 347–55.

49. Henry W. Castner, "Arthur Robinson: An Academic Family Tree," *Cartographic Perspectives* 51 (2005): 30–31.

50. Daniel R. Montello, "Cognitive Map-Design Research in the Twentieth Century: Theoretical and Empirical Approaches," *Cartography and Geographic Information Science* 29, no. 3 (2002): 283–304.

51. MacEachren articulates a response to the deconstructive trend in postmodern scholarship as it targets cartography: "What is needed, I believe, is a more balanced perspective on cartographic research that attempts to merge the perceptual, cognitive, and semiotic issues of maps as functional devices for portraying space and the sociocultural issues of how these portrayals might facilitate, guide, control, or stifle social interaction." And while he offers that he is no longer an adherent of the map communication model, the move toward map design research has its most significant roots in a functional map design moment. See Alan M. MacEachren, *How Maps Work: Representation, Visualization, and Design* (New York: Guilford Press, 2004), 11.

52. Montello, "Cognitive Map-Design Research," 298.

53. Mark Harrower, "The Cognitive Limits of Animated Maps," *Cartographica* 42, no. 4 (2007): 349.

54. MacEachren, *How Maps Work*, 280.

55. Indeed, MacEachren devotes several pages both to what we attend and where we attend on the map surface (80–101).

56. McCleary, "Discovering Cartography," 353. He continues, "When you stop looking at old maps from an archival perspective and focus on the fact that it is not the map which is most significant but rather the 'behavior' which produced the map and the 'behavior' which resulted from its use, then the role of any map or any group of maps becomes easy to understand."

57. Ibid., 281–87. Note that duration, rate of change, and order were three initial variables of dynamic maps as developed from MacEachren's team at Penn State; see David DiBiase, Alan M. MacEachren, John B. Krygier, and Catherine Reeves, "Animation and the Role of Map Design in Scientific Visualization," *Cartography and Geographic Information Systems* 19, no. 4 (1992): 201–14.

58. Mark Harrower and Sara Irina Fabrikant, "The Role of Map Animation for Geographic Visualization," in *Geographic Visualization: Concepts, Tools and Applications,* edited by Martin Dodge, Mary McDerby, and Martin Turner (New York: Wiley, 2008), 62.

59. Harrower, "Cognitive Limits of Animated Maps."

60. Carolyn Fish, Kirk P. Goldsberry, and Sarah Battersby, "Change Blindness in Animated Choropleth Maps: An Empirical Study," *Cartography and Geographic Information Science* 38, no. 4 (2011): 357.

61. Daniel Dorling and Stan Openshaw, "Using Computer Animation to Visualize Space-Time Patterns," *Environment and Planning B: Planning and Design* 19 (1992): 644.

62. MacEachren, *How Maps Work*, 90.
63. Jacques Rancière, *The Intervals of Cinema* (New York: Verso, 2014), 11.
64. Dorling, "Stretching Space and Splicing Time," 215.
65. Harrower and Fabrikant, "Role of Map Animation," 49.
66. Massey, *For Space*, 106–11.
67. Ibid.
68. Harrower, "Cognitive Limits of Animated Maps," 350.
69. Michel de Certeau, *The Practice of Everyday Life* (Berkeley: University of California Press, 1984), 89.
70. Massey, *For Space*, 111.

4. Attention

1. New graduate programs for making web-based cartography have emerged to channel the proliferation of new mapping industries, including my own program at the University of Kentucky and the program at the University of Wisconsin.

2. Compare then, N. Katherine Hayles, *How We Think: Digital Media and Contemporary Technogenesis* (Chicago: University of Chicago Press, 2012); and Bernard Stiegler, *Taking Care of Youth and the Generations* (Stanford, Calif.: Stanford University Press, 2010).

3. See Bernard Stiegler, *For a New Critique of Political Economy* (Malden, Mass.: Polity, 2010). Stiegler advances the concept of an economy of contribution.

4. Daniel Z. Sui and Michael F. Goodchild, "GIS as Media?," *International Journal of Geographic Information Science* 15, no. 5 (2001): 387–90; Daniel Z. Sui and Michael F. Goodchild, "The Convergence of GIS and Social Media: Challenges for GIScience," *International Journal of Geographical Information Science* 25, no. 11 (November 2011): 1737–48; Matthew W. Wilson and Monica Stephens, "GIS as Media?," in *Mediated Geographies and Geographies of Media*, edited by Susan Mains, Julie Cupples, and Chris Lukinbeal (New York: Springer, 2015), 209–21; Agnieszka Leszczynski, "Spatial Media/tion," *Progress in Human Geography* 39, no. 6 (2015): 729–51.

5. My focus on community-based organizations draws on the traditions of community-engaged GIS, including efforts in participatory and public participation GIS, as well as broader efforts in radical geography to affect change through scholarship. See William J. Craig, Trevor M. Harris, and Daniel Weiner, eds., *Community Participation and Geographic Information Systems* (New York: Taylor and Francis, 2002); Sarah A. Elwood, "Beyond Cooptation or Resistance: Urban Spatial Politics, Community Organizations, and GIS-Based Spatial Narratives," *Annals of the Association of American Geographers* 96, no. 2 (2006): 323–41.

6. Bunge's work with the Detroit Geographical Expedition and Institute (DGEI) and Elwood's work with the Humboldt Park GIS Project are but two examples of the commitment of geographers to utilize mapping and GIS technologies in support of community agendas. See William Bunge, *Fitzgerald: Geography of a Revolution* (Cambridge, Mass.: Schenkman, 1971); Sarah A. Elwood, "Negotiating Knowledge Production: The Everyday Inclusions, Exclusions, and Contradictions of Participatory GIS Research," *Professional Geographer* 58, no. 2 (2006): 197–208.

7. Nathan Van Camp, "From Biopower to Psychopower: Bernard Stiegler's Pharmacology of Mnemotechnologies," *CTheory* (2012); see also Bernard Stiegler, *What Makes Life Worth Living: On Pharmacology* (Malden, Mass.: Polity, 2013).

8. Here, I draw specifically on *Taking Care of Youth and the Generations*, but also see Bernard Stiegler, "Relational Ecology and the Digital Pharmakon," *Culture Machine* 13 (2012): 1–19.

9. Bernard Stiegler, *Uncontrollable Societies of Disaffected Individuals: Disbelief and Discredit* (Cambridge: Polity, 2012); Bernard Stiegler, *States of Shock: Stupidity and Knowledge in the Twenty-First Century* (Cambridge: Polity, 2015); Bernard Stiegler, *Symbolic Misery: The Catastrophe of the Sensible* (Malden, Mass.: Polity, 2015).

10. Sam Kinsley, "Memory Programmes: The Industrial Retention of Collective Life," *cultural geographies* 22, no. 1 (January 2015): 169. Here, Kinsley draws on work by Rob Kitchin and Martin Dodge in examining the advance of code; see Rob Kitchin and Martin Dodge, *Code/Space: Software and Everyday Life* (Cambridge, Mass.: MIT Press, 2011).

11. Stiegler, *Taking Care*, 13.

12. Along these lines, see also Catherine Malabou, *What Should We Do with Our Brain?* (New York: Fordham University Press, 2008).

13. Leszczynski, "Spatial Media/tion," 731. Here, she suggests that my advancement of a social history of GIS is to write "spatial media into GIS genealogies." Instead, I argue that the point of this project is to recognize moments that they become intertwined, such as at conferences like *Where 2.0*; see Matthew W. Wilson, "Location-Based Services, Conspicuous Mobility, and the Location-Aware Future," *Geoforum* 43, no. 6 (November 2012): 1266–75; Wilson and Stephens, "GIS as Media?"

14. Compare, for instance, the fervor of Stan Openshaw, "A View on the GIS Crisis in Geography, or, Using GIS to Put Humpty-Dumpty Back Together Again," *Environment and Planning A* 23, no. 5 (1991): 621–28; and the foundation perspective of Carl Steinitz, *A Framework for Geodesign: Changing Geography by Design* (Redlands, Calif.: Esri, 2012).

15. Matthew W. Wilson, "Towards a Genealogy of Qualitative GIS," in *Qualitative GIS: A Mixed Methods Approach*, edited by Meghan Cope and Sarah A. Elwood (London: Sage, 2009), 156–70.

16. Donna Haraway's development of the concept of the cyborg has been particularly useful in thinking through these partnerships and the impossibilities of reciprocity; see Matthew W. Wilson, "Cyborg Geographies: Towards Hybrid Epistemologies," *Gender, Place and Culture* 16, no. 5 (October 2009): 499–516.

17. Indeed, these forms of engagement often adopt multiple modes of media strategy; see Bryan Preston and Matthew W. Wilson, "Practicing GIS as Mixed-Method: Affordances and Limitations in an Urban Gardening Study," *Annals of the Association of American Geographers* 104, no. 3 (2014): 510–29.

18. As discussed in chapter 3 of this volume, "Movement," Arthur Robinson is the foundation for much of this thought in cartographic scholarship of the late twentieth century; see Arthur Robinson, *The Look of Maps: An Examination of Cartographic Design* (Madison: University of Wisconsin Press, 1952).

19. For an overview of the new landscape of online spatial media, see Jeremy W. Crampton, "Cartography: Maps 2.0," *Progress in Human Geography* 33, no. 1 (2009): 91–100. On the topic of the unevenness of experiences resulting from these changes in digital media, see Stephen Graham, "The End of Geography or the Explosion of Place? Conceptualizing Space, Place and Information Technology," *Progress in Human Geography* 22, no. 2 (1998): 165–85; Stephen Graham, "Software-Sorted Geographies," *Progress in Human Geography* 29, no. 5 (2005): 562–80; and Stephen Graham and Simon Marvin, *Splintering Urbanism: Networked Infrastructures, Technological Mobilities and the Urban Condition* (New York: Routledge, 2001).

20. API refers to application programming interface. For a discussion of the emergence of location-based services and geosocial media, see Matthew James Kelley, "The Emergent Urban Imaginaries of Geosocial Media," *GeoJournal* 78, no. 1 (2013): 181–203; Matthew W. Wilson, "Location-Based Services, Conspicuous Mobility, and the Location-Aware Future," *Geoforum* 43, no. 6 (November 2012): 1266–75.

21. Limor Shifman, "An Anatomy of a YouTube Meme," *New Media & Society* 14, no. 2 (2012): 187–203.

22. Consider the controversy associated with the "avoid ghetto" application; see Jim Thatcher, "Avoiding the Ghetto through Hope and Fear: An Analysis of Immanent Technology Using Ideal Types," *GeoJournal* 78, no. 6 (2013): 967–80.

23. Matthew W. Wilson, "Geospatial Technologies in the Location-Aware Future," *Journal of Transport Geography* 34 (2014): 297–99; Matthew W. Wilson, "Flashing Lights in the Quantified Self-City-Nation," *Regional Studies, Regional Science* 2, no. 1 (2015): 39–42.

24. Tim Carmody, "The Damning Backstory Behind 'Homeless Hotspots' at SXSW," *Wired*, March 12, 2012, www.wired.com/2012/03/the-damning-back story-behind-homeless-hotspots-at-sxswi/.

25. danah boyd and Kate Crawford, "Critical Questions for Big Data: Provocations for a Cultural, Technological, and Scholarly Phenomenon," *Information, Communication & Society* 15, no. 5 (2012): 662–79.

26. This "newness" has been the drumbeat of the GISciences, culminating with Mike Goodchild's article in 2007 on "citizens as sensors" and Andrew Turner's *Introduction to Neogeography;* see Michael F. Goodchild, "Citizens as Sensors: The World of Volunteered Geography," *GeoJournal* 69 (2007): 211–21; Andrew J. Turner, *Introduction to Neogeography, O'Reilly Short Cuts* (Sebastopol, Calif.: O'Reilly, 2006). Also see these edited collections: Sarah A. Elwood, "Volunteered Geographic Information: Key Questions, Concepts and Methods to Guide Emerging Research and Practice," *GeoJournal* 72 (2008): 133–35; Agnieszka Leszczynski and Matthew W. Wilson, "Theorizing the Geoweb," *GeoJournal* 78, no. 6 (July 2013): 915–19; Matthew W. Wilson and Mark Graham, "Situating Neogeography," *Environment and Planning A* 45, no. 1 (2013): 3–9; and Daniel Sui, Sarah A. Elwood, and Michael F. Goodchild, eds., *Crowdsourcing Geographic Knowledge: Volunteered Geographic Information (VGI) in Theory and Practice* (New York: Springer, 2013).

27. See Sarah A. Elwood and Agnieszka Leszczynski, "Privacy, Reconsidered: New Representations, Data Practices, and the Geoweb," *Geoforum* 42 (2011): 6–15; Nancy J. Obermeyer, "Thoughts on Volunteered (Geo)Slavery" (paper presented at Workshop on Volunteered Geographic Information, Santa Barbara, Calif., 2007).

28. Elwood and Leszczynski, "Privacy, Reconsidered," 13.

29. Agnieszka Leszczynski, "Situating the Geoweb in Political Economy," *Progress in Human Geography* 36, no. 1 (2012): 72–89; Thatcher, "Avoiding the Ghetto"; Wilson, "Location-Based Services"; Craig Dalton, "Sovereigns, Spooks, and Hackers: An Early History of Google Geo Services and Map Mashups," *Cartographica* 48, no. 4 (2013): 261–74.

30. Leszczynski, "Situating the Geoweb in Political Economy," 84.

31. For instance, Harvey's discussion of time-space compression with Scott Kirsch; see David Harvey, *The Condition of Postmodernity: An Enquiry into the Origins of Cultural Change* (Cambridge, Mass.: Blackwell, 1989); Scott Kirsch, "The Incredible Shrinking World? Technology and the Production of Space," *Environment and Planning D: Society and Space* 13 (1995): 529–55. The tradition of critical technology studies in geography is a relatively recent one; see Graham, "The End of Geography"; Kitchin and Dodge, *Code/Space;* Matthew A. Zook, *The Geography of the Internet Industry: Venture Capital, Dot-Coms, and Local Knowledge* (Malden, Mass.: Blackwell, 2005); Sam Kinsley, "Futures in the Making: Practices to Anticipate 'Ubiquitous Computing,'" *Environment and Planning A* 44 (2012): 1554–69; James Ash, "Attention, Videogames and the Retentional Economies of Affective Amplification," *Theory, Culture & Society* 29, no. 6 (2012): 3–26; Joe Gerlach, "Lines, Contours and Legends: Coordinates for Vernacular

Mapping," *Progress in Human Geography* 38, no. 1 (2014): 22–39; Nicholas Bauch, "Extensible, Not Relational: Finding Bodies in the Landscape of Electronic Information with Wireless Body Area Networks," *GeoJournal* 78 (2013): 921–34.

32. Hayles, *How We Think*; Stiegler, *Taking Care of Youth*; Malabou, *What Should We Do*; Patrick Crogan and Sam Kinsley, "Paying Attention: Towards a Critique of the Attention Economy," *Culture Machine* 13 (2012): 2–29.

33. Federica Frabetti, "Rethinking the Digital Humanities in the Context of Originary Technity," *Culture Machine* 12 (2011): 7.

34. Ibid.

35. James Ash, "Architectures of Affect: Anticipating and Manipulating the Event in Processes of Videogame Design and Testing," *Environment and Planning D: Society and Space* 28 (2010): 653–71; James Ash, "Teleplastic Technologies: Charting Practices of Orientation and Navigation in Videogaming," *Transactions of the IBG* 35, no. 3 (July 2010): 414–30; James Ash, "Technology, Technicity and Emerging Practices of Temporal Sensitivity in Videogames," *Environment and Planning A* 44 (2012): 187–203; Ash, "Attention, Videogames"; James Ash, "Rethinking Affective Atmospheres: Technology, Perturbation and Space Times of the Non-Human," *Geoforum* 49 (2013): 20–28; James Ash, *The Interface Envelope: Gaming, Technology, Power* (New York: Bloomsbury, 2015).

36. Sy Taffel, "Escaping Attention: Digital Media Hardware, Materiality and Ecological Cost," *Culture Machine* 13 (2012): 1–28.

37. Ben Roberts, "Attention Seeking: Technics, Publics and Free Software Individuation," *Culture Machine* 13 (2012): 1–20.

38. I borrowed this course title from a similar course taught by Sarah Elwood and Tim Nyerges at the University of Washington, the roots of which stretch into the late 1980s, when Nyerges and Chrisman designed the GIS and computer cartography curricula; see Timothy L. Nyerges and Nicholas R. Chrisman, "A Framework for Model Curricula Development in Cartography and Geographic Information Systems," *Professional Geographer* 41, no. 3 (1989): 283–93. My course has since gone through another redesign, and I have renamed the course Community Mapshop.

39. Read more about the Syracuse Community Geography Program at www.communitygeography.org. See also Timothy L. Nyerges, Michael Barndt, and Kerry Brooks, "Public Participation Geographic Information Systems" (paper presented at the AutoCarto 13, ACSM/ASPRS 1997 Technical Papers, Seattle, Washington, April 1996); Paul Schroeder, "Criteria for the Design of a GIS/2, Specialists' Meeting for NCGIA Initiative 19: GIS and Society," www.spatial.maine.edu/~schroedr/ppgis/criteria.html; Sarah A. Elwood, "Integrating Participatory Action Research and GIS Education: Negotiating Methodologies, Politics and Technologies," *Journal of Geography in Higher Education* 33, no. 1 (January 2009): 51–65.

40. To support this process, twelve interviews were conducted and transcribed with community partners during the summers of 2012 and 2013. Many thanks to Sonya Prasertong and Jessa Loomis, research assistants during this project.

41. Cheryl Truman, "Faith Feeds Provides Food and Helps Build a Community," *Lexington Herald-Leader,* July 25, 2012.

42. Interview with community partner, June 13, 2012.

43. Interview with community partner, June 7, 2012.

44. Interview with community partner, May 28, 2013.

45. Mark Graham, "Time Machines and Virtual Portals: The Spatialities of the Digital Divide," *Progress in Development Studies* 11, no. 3 (2011): 211–27; Melissa Gilbert and Michele Masucci, *Information and Communication Technology Geographies: Strategies for Bridging the Digital Divide* (Vancouver: Praxis Press, 2011); Barney Warf, "Segueways into Cyberspace: Multiple Geographies of the Digital Divide," *Environment and Planning B: Planning and Design* 28, no. 1 (2001): 3–19.

46. Interview with community partner, June 6, 2012.

47. Ibid.

48. Ibid.

49. Interview with community partner, May 31, 2013.

50. Interview with community partner, June 6, 2012.

51. Available at www.mailchimp.com, MailChimp advertises itself as an "Online email marketing solution to manage contacts, send email and track results."

52. Interview with community partner, June 6, 2012.

53. Interview with community partner, June 4, 2013.

54. Interview with community partner, May 28, 2013.

55. Although, see Stiegler's discussion of what he calls the economy of contribution; Bernard Stiegler, *The Re-Enchantment of the World: The Value of Spirit against Industrial Populism* (London: Bloomsbury, 2014).

56. See Sarah A. Elwood and Helga Leitner, "GIS and Spatial Knowledge Production for Neighborhood Revitalization: Negotiating State Priorities and Neighborhood Values," *Journal of Urban Affairs* 25, no. 2 (May 2003): 139; Rina Ghose, "Use of Information Technology for Community Empowerment: Transforming Geographic Information Systems into Community Information Systems," *Transactions in GIS* 5, no. 2 (2001): 141–63.

57. Brian Massumi, *The Power at the End of the Economy* (Durham, N.C.: Duke University Press, 2015).

58. Jacques Derrida, *Of Grammatology* (Baltimore, Md.: Johns Hopkins University Press, 1997), 292.

59. William E. Connolly, *Neuropolitics: Thinking, Culture, Speed* (Minneapolis: University of Minnesota Press, 2002), 104.

60. Stiegler, *For a New Critique,* 103.

61. Stiegler, *Taking Care,* 227–28n13.

62. Ibid., 185.

63. Arthur Howard Robinson, *Elements of Cartography*, 2nd ed. (New York: Wiley, 1960), 14. Interestingly, my eye was drawn to this marked quote in a copy of this book that was stamped by the Harvard Laboratory for Computer Graphics on September 14, 1966. I would like to think the annotations of the text were Howard Fisher's, however improbable this might be. By 1966 Fisher was in deep conversation regarding the psychological aspects of digital maps.

64. For instance, the closing statement of Fish et al. in 2011: "More research is needed to better understand these issues and to inform future designs that can overcome or minimize their effects." See Carolyn Fish, Kirk P. Goldsberry, and Sarah Battersby, "Change Blindness in Animated Choropleth Maps: An Empirical Study," *Cartography and Geographic Information Science* 38, no. 4 (2011): 350–62.

65. Robinson, *The Look of Maps*, 48.

5. Quantification

1. Félix Ravaisson, *Of Habit* (New York: Continuum, 2008), 27. He continues, "The conditions under which being is represented to us in the world are Space and Time."

2. Richard Sennett, *Flesh and Stone: The Body and the City in Western Civilization* (New York: W. W. Norton, 1994).

3. Michael Batty, *The New Science of Cities* (Cambridge, Mass.: MIT Press, 2013).

4. Elvin K. Wyly, "The New Quantitative Revolution," *Dialogues in Human Geography* 4, no. 1 (2014): 26–38; also see Mark Graham and Taylor Shelton, "Geography and the Future of Big Data, Big Data and the Future of Geography," *Dialogues in Human Geography* 3, no. 3 (2013): 255–61.

5. See Doreen J. Mattingly and Karen Falconer-Al-Hindi, "Should Women Count? A Context for the Debate," *Professional Geographer* 47, no. 4 (1995): 433, where they conclude, "Most qualitative research involves some kind of counting, and that quantitative methods involve an array of interpretive acts."

6. Louise Amoore, "Algorithmic War: Everyday Geographies of the War on Terror," *Antipode* 41, no. 1 (2009): 64.

7. Donna Haraway is my guide, here; she writes of her approach: "this essay is a contribution to the heterogeneous and very lively contemporary discourse of science studies as cultural studies"; see Donna Haraway, "The Promises of Monsters: A Regenerative Politics for Inappropriate/d Others," in *Cultural Studies*, edited by Lawrence Grossberg, Cary Nelson, and Paula A. Treichler (New York: Routledge, 1992), 296.

8. As Foucault argued in "Theatrum Philosophicum," *Critique* 282 (1970): 885–908, "What is the answer to the question? The problem. How is the problem resolved? By displacing the question."

9. See www.ibm.com/smarterplanet/us/en/smarter_cities/overview/ and https://jawbone.com/up.

10. Parmy Olson, "A Massive Social Experiment on You Is Under Way, and You Will Love It," *Forbes*, February 9, 2015, www.forbes.com/sites/parmyolson/2015/01/21/jawbone-guinea-pig-economy/.

11. Gary Wolf and Kevin Kelly, "Quantified Self," http://quantifiedself.com/about/.

12. Boston Quantified Self was founded in January 2010, boasting nearly fifteen hundred members, according to their Meetup webpage, www.meetup.com/BostonQS/.

13. Claire Elaine Rasmussen, *The Autonomous Animal: Self-Governance and the Modern Subject* (Minneapolis: University of Minnesota Press, 2011).

14. Michel de Certeau, *The Practice of Everyday Life* (Berkeley: University of California Press, 1984), 129.

15. Ibid., 121.

16. Henri Lefebvre, *The Production of Space* (Cambridge, Mass.: Blackwell, 1991), 7.

17. John K. Wright, "Map Makers Are Human: Comments on the Subjective in Maps," *Geographical Review* 32, no. 4 (1942): 544.

18. Nadine Schuurman, "Social Perspectives on Semantic Interoperability: Constraints on Geographical Knowledge from a Data Perspective," *Cartographica* 40, no. 4 (2005): 49.

19. For instance, see the burgeoning literature surrounding the quantification of nature, the city, and the body: Eric Nost, "Performing Nature's Value: Software and the Making of Oregon's Ecosystem Services Markets," *Environment and Planning A* 47 (2015): 2573–90; Venkata Krishna Kumar Matturi, "Smart Urbanization: Emerging Paradigms of Sensing and Managing Urban Systems," *Planum: The Journal of Urbanism* 27, no. 2 (2013): 1–8; Taylor Shelton, Matthew A. Zook, and Alan Wiig, "The 'Actually Existing Smart City,'" *Cambridge Journal of Regions, Economy and Society* 8 (2015): 13–25; Kate Crawford, "When Big Data Marketing Becomes Stalking," *Scientific American*, January 28, 2014, www.scientificamerican.com/article/when-big-data-marketing-becomes-stalking/.

20. Batty, *New Science*.

21. Michael Batty, "Smart Cities, Big Data," *Environment and Planning B: Planning and Design* 39 (2012): 193.

22. Batty's new science could be understood more as a continuation of many of the motivations of social physics and spatial analysis; see Trevor J. Barnes and Matthew W. Wilson, "Big Data, Social Physics, and Spatial Analysis: The Early Years," *Big Data & Society* 1, no. 1 (April–June 2014): 1–14.

23. Batty, *New Science*, 47.

24. Ibid., 45.

25. Anthony M. Townsend, *Smart Cities: Big Data, Civic Hackers, and the Quest for a New Utopia* (New York: W. W. Norton, 2013), 14.

26. Ibid., 15.

27. Ibid., 284–85.

28. Indeed, much of this critique is anticipated by Geoffrey C. Bowker and Susan Leigh Star, *Sorting Things Out: Classification and Its Consequences* (Cambridge, Mass.: MIT Press, 1999). See also Jim Thatcher, David O'Sullivan, and Dillon Mahmoudi, "Data Colonialism through Accumulation by Dispossession: New Metaphors for Daily Data," *Environment and Planning D: Society and Space* 34, no. 6 (2016): 990–1006.

29. Batty, *New Science*, 47.

30. Sennett, *Flesh and Stone*.

31. Jordan Crandall, "The Geospatialization of Calculative Operations: Tracking, Sensing and Megacities," *Theory, Culture & Society* 27, no. 6 (2010): 72.

32. Sennett, *Flesh and Stone*, 275.

33. See Matthew W. Wilson, "Location-Based Services, Conspicuous Mobility, and the Location-Aware Future," *Geoforum* 43, no. 6 (November 2012): 1266–75; Matthew W. Wilson, "Geospatial Technologies in the Location-Aware Future," *Journal of Transport Geography* 34 (2014): 297–99; Matthew W. Wilson, "Flashing Lights in the Quantified Self-City-Nation," *Regional Studies, Regional Science* 2, no. 1 (2015): 39–42; also Jim Thatcher, "Avoiding the Ghetto through Hope and Fear: An Analysis of Immanent Technology Using Ideal Types," *GeoJournal* 78, no. 6 (2013): 967–80; Matthew James Kelley, "The Emergent Urban Imaginaries of Geosocial Media," *GeoJournal* 78, no. 1 (February 2013): 181–203; Matthew James Kelley, "Urban Experience Takes an Informational Turn: Mobile Internet Usage and the Unevenness of Geosocial Activity," *GeoJournal* 79, no. 1 (February 2014): 1529.

34. Martin Dodge and Rob Kitchin explore surveillance and sousveillance in the design of shared public toilets; Martin Dodge and Rob Kitchin, "Towards Touch-Free Spaces: Sensors, Software and the Automatic Production of Shared Public Toilets," in *Touching Space, Placing Touch*, edited by Mark Paterson and Martin Dodge (Burlington, Vt.: Ashgate, 2012), 191–210.

35. The selfie is a photograph of oneself, sometimes taken with others, often taken to be an expression of self-in-place, a kind of conspicuous mobility. The commonplace of this action has resulted in digital devices like the iPhone actually recognizing selfies from other photographs, as well as the global commodity of the selfie stick, which allows an individual to take a selfie without the visible appearance of an arm held out at a distance, to increase the conspicuousness of the place.

36. Amoore, "Algorithmic War," 64.

37. Crandall, "Geospatialization of Calculative Operations," 73.

38. See advertisement for the Jawbone UP personal activity monitor in Figure 22.

39. See advertisements for mobile applications called Automatic, www.automatic.com/, and Spreadsheets, http://spreadsheetsapp.com/.

40. For instance, see the design practice of Chris Speed, http://chrisspeed.net/.

41. Notably see Amoore, "Algorithmic War"; Ben Anderson, "Preemption, Precaution, Preparedness: Anticipatory Action and Future Geographies," *Progress in Human Geography* 34, no. 6 (2010): 777–98; Jeremy Crampton, Susan Roberts, and Ate Poorthuis, "The New Political Economy of Geographical Intelligence," *Annals of the Association of American Geographers* 104, no. 1 (2014): 196–214.

42. For instance, the series of popular horror films under the title of *Paranormal Activity* documents a society increasingly interested and capable of looking in on itself. The films proceed through amateur video recordings and surveillance systems set up in homes. Without these technologies, the films and the fear they engender in the audience would simply not be possible. I suggest that the big data society is a contemporary incarnation of such a specter. Therefore, the uneasiness that forms this franchise is not just another ghost story, but the increasing insecurity many feel within the spaces of their homes, their social media, their countries, and their bodies, and the necessary technical solutions to document and further establish these senses of insecurity and vulnerability.

43. See Rio's hosting of the World Cup in 2014 and the symbolic importance of their smart city operations; Christopher Frey, "World Cup 2014: Inside Rio's Bond-Villain Mission Control," *Guardian*, May 23, 2014, www.theguardian.com/cities/2014/may/23/world-cup-inside-rio-bond-villain-mission-control.

44. See Federico Guerrini, "World's Top 7 Smart Cities of 2015 Are Not the Ones You'd Expect," *Forbes*, January 28, 2015, www.forbes.com/sites/federicoguerrini/2015/01/28/worlds-top-7-smartest-cities-of-2015-are-not-the-ones-youd-expect/print/.

45. For instance, see Stephen Graham, *Disrupted Cities: When Infrastructure Fails* (New York: Routledge, 2010); and Rob Kitchin and Martin Dodge's book *Code/Space*, which outlines situations in which the functionality of designed spaces is completely derailed when code fails: Rob Kitchin and Martin Dodge, *Code/Space: Software and Everyday Life* (Cambridge, Mass.: MIT Press, 2011).

46. Fitbit statement in Timothy Stenovec, "Some Fitbit Force Owners Complain of Severe Skin Irritation," *Huffington Post,* January 23, 2014, www.huffingtonpost.com/2014/01/13/fitbit-force-rashes_n_4590859.html.

47. See Natasha Singer, "Mission Control, Built for Cities," *New York Times*, March 3, 2012.

48. Gunnar Olsson, *Abysmal: A Critique of Cartographic Reason* (Chicago: University of Chicago Press, 2007), 10.

49. John Sallis, *Klee's Mirror* (Albany: State University of New York Press, 2015), 55.

50. Klee quoted from his 1921 Bauhaus lectures, in ibid., 56.

6. A Single Point Does Not Form a Line

1. Derek Gregory, "Editorial: Gregory D. (1990)," *Environment and Planning D: Society and Space* 8 (1990): 1–6. Many thanks to Ate Poorthuis for pointing me to this editorial, during our discussion of the use of footnotes in academic writing.
2. Bernard Stiegler, "Relational Ecology and the Digital Pharmakon," *Culture Machine* 13 (2012): 2.
3. Ibid., 1.
4. Arthur Howard Robinson, *Elements of Cartography,* 2nd ed. (New York: Wiley, 1960), 15.
5. One of MacEachren's instructors, George McCleary, also recognized that key questions were not being asked within the traditions of cartography: "Not all cartographic activity, however, was concerned with applying the principles and procedures of the mapmaker's craft"; see George McCleary, "Discovering Cartography as a Behavioral Science," *Journal of Environmental Psychology* 7 (1987): 348.
6. Tom Conley, "Mapping in the Folds: Deleuze *Cartographe,*" *Discourse* 20, no. 3 (1998): 134.
7. See National Science Foundation, "Science and Engineering Indicators," edited by National Center for Science and Engineering Statistics, Arlington, Va., 2014, www.nsf.gov/statistics/seind14/index.cfm/chapter-7; Scott Neuman, "1 in 4 Americans Think the Sun Goes around the Earth, Survey Says," *National Public Radio,* February 14, 2014, www.npr.org/sections/thetwo-way/2014/02/14/277058739/1-in-4-americans-think-the-sun-goes-around-the-earth-survey-says.
8. Isaiah Bowman, *Geography in Relation to the Social Sciences* (New York: Charles Scribner's Sons, 1934), 20.
9. See chapter 2; Neil Smith, "'Academic War over the Field of Geography': The Elimination of Geography at Harvard, 1947–1951," *Annals of the Association of American Geographers* 77, no. 2 (1987): 155–72.
10. Guy Debord, *The Society of the Spectacle* (New York: Zone Books, 1994).
11. Guy Debord, "Introduction to a Critique of Urban Geography," in *Critical Geographies: A Collection of Readings,* edited by Harald Bauder and Salvatore Engel-Di Mauro (Kelowna, BC: Praxis, 2008), 25.
12. To quote one anonymous reviewer, the slow map "seems hopelessly inadequate, idealistic, and miniscule compared to the huge forces of globalized capital and the digitization of the earth and all its inhabitants." I suggest that this is the point. In the fast-paced world of global capitalism, such attendance to the miniscule can seem nostalgic and hopelessly romantic; yet, an address is necessary that does not simply add fuel to the flame but contemplates a different rhythm and a different vision.
13. NPR memo to staff, April 28, 1994.

14. The notion of "continuous connectivity" through mobile devices grows to a fever pitch on the backs of the personal digital assistant in the late 1990s; see Matthew W. Wilson, "Continuous Connectivity, Handheld Computers, and Mobile Spatial Knowledge," *Environment and Planning D: Society and Space* 32, no. 3 (2014): 538.

15. Arthur Robinson, *The Look of Maps* (Madison: University of Wisconsin Press, 1952), 73.

16. Gilles Deleuze, *Cinema 1: The Movement-Image*, translated by Hugh Tomlinson and Barbara Habberjam (London: Bloomsbury Academic, 2013), 194–95.

17. Ibid., 168.

INDEX

academy, ix, 1–3, 5, 10, 18–22, 26, 27, 39, 49, 59, 63, 67, 98–99, 113, 136, 139, 146n1
advertising, 11, 33, 34, 48–49, 75–76, 80, 83, 88, 95, 97, 114, 116, 119, 120, 121, 128–29, 164n19. *See also* marketing
Amoore, Louise, 117, 130
animated cartography, 23, 58, 71, 79, 82, 83–86, 87, 127, 164n35. *See also* animated map
animated map, 71–74, 76, 78, 79, 80, 81–93. *See also* animated cartography
Apple, ix, 100, 118
ArcGIS, 4, 14, 34, 35. *See also* Esri
art, 37, 38, 44, 47, 49, 75, 81, 82–86, 89, 91, 93, 112, 140, 141
Ash, James, 102–3
AT&T, 128–29

Battersby, Sarah, 89
Batty, Mike, 115–16, 124–25, 127, 174n22
Bauhaus, x, 176n50
behavioral cartography, 79, 86–88, 90, 93, 114. *See also* behavioral map research

behavioral map research, 16, 141, 165n47, 166n56. *See also* behavioral cartography
Benson, Betty, 51, 56, 66, 157n15, 159n34
Bergson, Henri, 71, 78, 80
Berry, Brian, 66
Bertin, Jacques, 84, 86, 87
Bieber, Justin, 138
big data, 10–13, 33–34, 70, 77, 124, 153n35, 176n42
body, ix, 22, 23, 42, 112, 115, 116, 118–21, 124, 126, 127–28, 130–32, 140, 174n19
Body of Knowledge: AAG and UCGIS, 39–40, 70, 163n3
Bowman, Isaiah, 41, 53, 137–38
Bowman Expeditions, 41, 152n24
Brenner, Neil, 38. *See also* Urban Theory Lab
Brewer, Cindy, 86
Bunge, William, 27, 38, 43, 47, 54–55, 96, 168n6

care, 29, 33, 51, 96, 113, 115, 121, 122, 131, 135–36, 138–39, 155n62
CARTO, 34, 48, 73, 74, 79

Center for Geographic Analysis, 159n40, 161n59
Chagoya, Enrique, 37
Chrisman, Nicholas, viii, 31, 171n38
cinema: as condition, 69, 73, 91, 140, 165n36, 176n42; as maps, 38, 70, 82, 93, 127; as memory, 71, 80; as movement, 71, 80, 81
citizen science, 5, 30
click bait, 138
color, 16, 63, 66, 86, 87, 114
community-based, 5, 6, 12, 15, 16, 18, 21, 23, 31, 96–99, 102–13, 167n5, 168n6, 171n39. *See also* participatory; public participation
Conant, Jim, 53, 138, 158n24
Conley, Tom, 136
Connolly, William, 112
Corner, James, 32–33
correspondence course, 58, 62, 159n39. *See also* massively open and online courses
counting, 60, 116, 118, 121, 122, 131–32, 173n5
Crampton, Jeremy, 6, 147n10
Crandall, Jordan, 127, 130
critical cartography, 3, 7, 13, 14, 18, 27, 38, 66, 93, 133
Critical Cartography Collective, 38
critical GIS: agenda, xi, 4–5, 10, 31, 98–100, 102, 104; and crisis of representation, 27, 45; definition, 9, 26; origins of, x, 3–4, 28, 30, 124, 151n10; pedagogy, 12, 18–19, 39, 41, 96, 103–4, 108, 140; practice, 101. *See also* community-based; feminist GIS; qualitative GIS
cyberinfrastructure, 10
cyborg, 169n16

Dangermond, Jack, 58
Davis, William Morris, 52, 157

Debord, Guy, 26–27, 31, 38, 138–39
de Certeau, Michel, 93, 122
decision making, 34–35, 57, 89, 108, 119–21, 123, 128, 165n47
Deleuze, Gilles, vii, 7, 13, 14, 17, 47, 69–73, 78, 80–82, 85, 86, 140, 164n35, 165n35
Derrida, Jacques, 111–12
design, x, 2, 15, 19, 21, 23, 32–33, 35, 38, 43–45, 71, 75–76, 81, 84, 86–94, 95–96, 106, 113–14, 116, 124, 126, 132, 136, 139, 153n35, 164n19, 173n64, 175n34, 176n40
DiBiase, David, 39–40, 86, 88
digital culture, 10, 31, 39, 96, 99, 101, 105–6, 109, 112–13
digital geographies, 3, 5
digital humanities, 3, 5, 9, 10, 30–31, 102
Dorling, Danny, 79–80, 90, 91
drone, 20
Dutton, Geoff, 23, 63, 71, 90–92, 160n55

Elwood, Sarah, ix, 96, 101, 168n6, 171n38
Esri, 4, 19, 34–35, 39, 48, 58, 77, 79, 119, 126. *See also* ArcGIS
ethics, 8–9, 31, 39, 41, 70, 163n3

Fabrikant, Sara, 89, 92–93
Facebook, 105–7, 109–11, 124
Falconer-Al-Hindi, Karen, 116
feminist GIS, 9, 12, 30
Field, Kenneth, 77, 79, 80
Fish, Carolyn, 89, 173n64
Fisher, Howard, 22–23, 31, 42–44, 47, 49–52, 56–67, 86, 87, 159n34, 160n44, 160nn55–56, 163n9, 173n63
Fitbit, 117, 119, 132

Ford Foundation, 57, 58, 59, 63, 66, 160n56
Foursquare, 127
Frabetti, Federica, 102
Friday Harbor meetings, 8, 12, 28–32, 151n9

Garrison, Bill, 54, 55, 59, 66
General Houses, Inc., 51, 52, 63
geodesign, 33, 34–35
geohumanities, 10, 30–31
geointelligence, 5, 10, 23. See also Bowman Expeditions; geospatial intelligence
geospatial intelligence, 31. See also geointelligence
geovisualization, 5, 44
geoweb, 9, 30, 101, 154n49
GIS: expanded definition of, vii–ix, 4, 98; industry, 23, 31; 1990s debates on, 3, 10, 12, 13, 19, 22, 29, 90; pedagogy, 21, 39, 70, 101, 103–5, 113; as science, ix, 6, 8, 10, 12, 22, 35, 96, 98, 100–101, 124, 136, 147n7, 150n50; social history of, 7, 8, 10, 12, 13, 31, 168n13; and society, x, xi, 3, 8, 9, 12, 28, 30, 102, 111, 136. See also critical GIS; feminist GIS; qualitative GIS
Goldsberry, Kirk, 89, 173n64
Google, viii, 4, 10, 11, 48, 76, 100, 107, 109, 146n1
Graduate School of Design, 22–23, 38, 51, 58, 66, 90, 159n38
Graham, Stephen, ix
Gregory, Derek, 135
Guattari, Félix, vii, 7, 13, 17

habit, xi, 6, 14, 24, 27, 101, 112, 115–18, 120, 123, 127–30, 132, 137, 139, 145n5
Hägerstrand, 35

Haraway, Donna, vii, 7–8, 28, 49, 74, 145n3, 169n16, 173n7
Harley, Brian, vii, 14, 31, 145n1
Harrison, Richard Edes, 31, 44, 155n62, 163n11
Harrower, Mark, 87, 89, 92–93
Harvard University, 22–23, 38, 42–43, 47, 49, 50–54, 56–60, 63–64, 67, 69, 73, 85, 86, 90–92, 137–38, 157nn17–18, 159n34, 159n40, 160n55, 162n72, 173n63
Hayles, N. Katherine, 35–36, 95–96, 112, 154n41
Horwood, Edgar M., 55–56, 58, 65–66, 159n34

IBM, 61, 119, 121, 130, 132, 163n9
Ingold, Tim, 149n36
Institute of Geographical Exploration, 52–53, 157n18, 158n25
interoperability, 4, 36, 117, 119, 123–26, 130, 132
Isard, Walter, 69–70

Jawbone, 120, 121
Jenks, George, 51, 59, 65, 164n18

Kardashians, 20
Kinsley, Sam, 97
Kitchin, Rob, ix
Klee, Paul, 133–34
Krinke, Rebecca, 16–17
Kwan, Mei-Po, 10

labor, 21, 34, 42, 107
Laboratory for Computer Graphics, 23, 43, 47, 49, 57–58, 60, 66, 69, 86, 91–92, 156n3, 157n15, 159n38, 162n72, 173n63
Latour, Bruno, vii, 26, 39, 145n2, 155n51
Lefebvre, Henri, x, 123

Leszczynski, Agnieszka, 6, 98, 101,
  147n10, 149n41, 152n26, 168n13
location-aware, viii, x, 28, 34–35, 39,
  76, 113, 116, 117, 119, 123, 131,
  136, 138, 140. *See also* location-
  based; location-enabled
location-based, 4, 5, 10, 23, 34, 96,
  100, 102, 128–29, 131, 169n20.
  See also location-aware;
  location-enabled
location-enabled, 34. *See also*
  location-aware; location-based
Lynch, Kevin, 31

MacEachren, Alan, 86–88, 90,
  136, 166n51, 166n55, 166n57,
  177n5
MailChimp, 108–9, 111, 172n51
Mango, 48
Mapbox, 34, 48
map communication model, 44, 71,
  77, 87, 89, 90, 166n51
map story, 34. *See also* spatial history;
  story map
marketing, 26, 33–34, 48–49, 75–76,
  88, 97, 100, 108–10, 116, 119,
  128–29. *See also* advertising
Massey, Doreen, 36, 77–80, 93,
  154n41
massively open and online courses, 19
Mattingly, Doreen, 116
McCleary, George, Jr., 88, 166n56,
  177n5
memory, 8, 14, 33, 71, 78, 80, 96–97,
  101, 102, 111–13, 136–39
Moellering, Hal, 85–86, 165n47
Mogel, Lize, 38
Monmonier, Mark, 66
Montello, Dan, 87

National Center for Geographic
  Information and Analysis, 8

neogeography, 10, 30, 31, 33, 34, 40,
  170n26
neo-Robinsonian, 15, 76–77, 78, 80,
  86, 114, 139. *See also* Robinson,
  Arthur
nonrepresentational, 9
North American Cartographic
  Information Society, 139
Nyerges, Timothy, 165n47, 171n38

Obermeyer, Nancy, 101
Olsson, Gunnar, xi, 5–6, 7, 13, 133,
  147n9
Openshaw, Stan, 23, 29, 90, 150n50,
  168n14
O'Sullivan, David, 10
OXAV, 57

Parker, Edith Putnam, 31
participatory: action, 3, 9; engagement,
  102; GIS, 18, 30, 103, 167n5;
  mapping, 16–18, 40–41, 96,
  155n52. *See also* community-based;
  public participation
Patriot Act, 131
Pavlovskaya, Marianna, 25
Penn State University, 19, 39, 86, 88,
  166n57
personal activity monitors, 23, 117,
  118, 120–23, 126, 130–32. *See also*
  quantified self
pharmaka, 33, 111–12, 117, 133
Pickles, John, 8, 13
planetary urbanization, 38, 127, 131
Plato, 111–12
postcritical, 26, 151n2. *See also*
  Pruchnic, Jeff
Pruchnic, Jeff, 28
psychogeography, 16–17, 26. *See also*
  Debord, Guy
psychopower, 33, 93, 95, 97, 111–12,
  123, 130–32

public participation, 18, 30, 103, 110, 167n5. *See also* community-based; participatory

public scholarship, ix, 1, 5, 21–22, 26

qualitative GIS, 5, 9, 12, 25–26, 30, 36–37, 70–71, 147n12

quantified self, 33, 34, 121, 130. *See also* personal activity monitors

Raisz, Erwin, 31, 42, 43, 44, 49, 52–53, 74, 140, 155n57, 157n18, 158n20, 158n25

Rasmussen, Claire, 121

Ravaisson, Félix, 115, 173n1

retention, 32–33, 96–98, 111–14, 118

rhizomology, vii, xii, 7, 13, 14, 140

Rice, Alexander Hamilton, 52–53

Roberts, Ben, 103

Robinson, Anthony, 19

Robinson, Arthur, 14–15, 51, 59–60, 65, 66, 71, 74–76, 80, 81, 83, 84, 86, 87, 89, 95, 113–14, 136, 158n25, 163n11, 164nn19–20. *See also* neo-Robinsonian

Schaefer, Fred, 27, 47, 117, 156n2

Schmidt, Allan, 58, 65, 66, 71, 72–73, 85, 159n39, 163n9

Schulten, Susan, 44

Schuurman, Nadine, 10, 29, 30, 124, 151n10

science and technology studies, 5

selfie, 130, 175n35

Sennett, Richard, 115–16, 127–28, 131–32

Shepard, Don, 58, 63, 159n34

Sheppard, Eric, 30, 31

Sherman, John, 65

slow-mapping, 45, 70, 139–40, 177n12

smart city, 5, 23, 116–21, 123, 125–26, 130–33, 176n43

Smith, Neil, 14, 21–22, 30, 53

social media, 23, 98, 100–102, 105–7, 132, 138, 176n42

social physics, 49, 58, 90, 124, 174n22

spatial history, 36–38, 154n41, 162n2

spatial media, 33, 98–100, 102, 104, 113, 118, 168n13, 169n19

Steinitz, Carl, 65, 161n64

Stiegler, Bernard, 32, 33, 95–97, 102–3, 111, 113, 117, 118, 135–36, 139

story map, 34. *See also* map story; spatial history

surveil, 27, 31, 117, 123, 130–32, 175n34, 176n42

Synagraphic Mapping System, 51, 56–63, 65–66, 73, 85, 159n34, 163n9

Syracuse Community Geography, 103, 171n39

Taffel, Sy, 103

Taylor, Peter, 29

technoculture, viii–x, 4, 5, 32, 100, 117–19, 121, 123, 145n5

technopositionality, 9–10, 13, 15, 29, 38–41, 99, 148n19, 151n13, 154n48

technosocial, 34, 38, 40, 50, 80, 124

Thayer, Tom, 64–65, 161n64

thematic mapping, 43, 51, 60, 65, 67, 84, 86

Thrower, Norman, 83

time: and being, 115; as duration, 78, 88–89, 103, 130, 166n57; liveliness, 31, 35–38, 39, 77, 93, 127; representation of, 153n39; as speed, 70; as study of process, 69–70, 137. *See also* spatial history

Tobler, Waldo, 23, 54, 55, 59, 65, 66, 71, 85, 89

Torp, Tanya, 105

Townsend, Anthony, 125–26

Twitter, 70, 73–74, 79, 105–7, 109–11, 153n35

Ullman, Edward, 54
University Consortium for GIScience, 39–40, 163n3
University of Washington, 8, 54–55, 59, 65, 86, 165n47, 171n38. *See also* Friday Harbor meetings
Urban Theory Lab, 38, 127
USA Freedom Act, 131

Vietnam War, 64–65, 161nn64–65
volunteered geographic information, 30, 31, 170n26. *See also* neogeography
Von Thunen, 124

Warntz, William, 23, 27, 47, 58, 62–63, 66–67, 69, 90
Weinberg, Robert, 58, 66
Whittlesey, Derwent, 53, 58
Widener, Eleanor, 52
Williams, Robert, 61, 67
Woldenberg, Michael, 90–91
Wright, J. K., 31, 123, 163n11
Wyly, Elvin, 116

Zipf, George, 124. *See also* social physics
Zuckerberg, Mark, 124. *See also* Facebook

MATTHEW W. WILSON is associate professor of geography at the University of Kentucky and visiting scholar at the Center for Geographic Analysis at Harvard University. He cofounded and codirects the New Mappings Collaboratory, which studies and facilitates new engagements with geographic representation. His research in critical GIS draws on science and technology studies and urban political geography to understand the development and proliferation of location-based technologies within a pervasive digital culture.